# AN Archaeological Perspective OF Human History AT Grand Canyon

Christopher M. Coder

GRAND CANYON CONSERVANCY
GRAND CANYON, ARIZONA

*Dedicated to everyone who helps to take care of Grand Canyon National Park.*

GRAND CANYON CONSERVANCY

P.O. Box 399, Grand Canyon, AZ 86023
(800) 858-2808 grandcanyon.org



Composed in the United States of America
Printed in the United States of America
Second Edition 2021

ISBN 978-1-934656-14-3

Edited by Louella J. Holter
Designed by Rudy Ramos

Grand Canyon Conservancy is the official nonprofit partner of Grand Canyon National Park, raising private funds, operating retail shops within the park, and providing premier guided educational programs about the natural and cultural history of the region. Our supporters fund projects including trails and historic building preservation, educational programs for the public, and the protection of wildlife and their natural habitat.

Grand Canyon Conservancy inspires people to protect and enhance Grand Canyon National Park for present and future generations.

# TABLE OF CONTENTS

*The well-trained archaeologist working in Grand Canyon, or anywhere else for that matter, must first and foremost be an anthropologist.*

—Robert C. Euler

# INTRODUCTION

For millions of visitors coming to the Grand Canyon each year, the view surpasses even the jaded expectations of modern times. Seeing the place for just a few moments is all it takes to get the canyon in your blood. Usually less than a 24-hour stay for most people, it can be a sine wave of rushing and then waiting. Whether viewed from the rim down, the bottom up, or somewhere in between, *the overwhelming presence of the landscape* never fails to make an impression and the images linger with us. The experience, like salvation, is better felt than told.

For millennia, the canyon was home to American Indians, encompassing all of its visual beauty, simple pleasures, and social complexities; it was hearth, school, hardware and grocery store, museum and church. The evidence in the form of 11,000-year-old spear points, Archaic hunting camps, and worn-down Puebloan villages is spread out across the landscape like weathered veneer, rarely obvious and mostly erased by the relentless processes of time.

Grand Canyon and the region it defines encompasses thousands of square miles of incised canyons, isolated mesas, and dormant volcanoes overlain by a spider web of drainages that connect Grand Canyon proper to the world outside, easing the movement of people into and through a region that was otherwise all but inaccessible. Before 1869 no fences or invisible lines crossed the countryside; the land just blended into upriver, downriver, and the world beyond the rims. In 1953 Mike Tobin, a Chemehuevi man, when pressed by lawyers of the Indian Claims Commission about the old borders between his people and the Hualapai in

The Grand Canyon region has been home to humans for more than 13,000 years.

his father's time, said simply, "They didn't know nothing about the lines." Lines like that were not a part of the tribal world.

Convolutions in the rock, caves, cracks beyond count, the connective tissue of scores of named canyons and hundreds with no name, combine to create an unimaginable space within the canyon. Grand Canyon can be measured, geological formations scrutinized, pottery and stone tools analyzed, sites mapped to scale, but quantifying the whole is elusive if not impossible. There remains something intangible about the experience, which archaeologists—as observers from another age with a different perspective—cannot quite get their arms around. However, ongoing work, collaboration with tribal elders, and ever-better science help refine and redefine our understanding with each passing decade.

The notion of time is a human convention designed to track change and motion. People's lives and the cultures they create move through the years as braided streams, leaving ever-changing trails of evidence in their wakes. Some of the evidence clarifies; some of it confuses or begs for more. The goods we create and the clues left behind are all subtle indications of the unique behaviors that set us apart from our neighboring cultures in time and in place. Birth and puberty rites, style of burial, style of art, style of warfare, form of worship, food taboos, the games we play (or watch) are all lines on the changing face of culture. As individuals, we are not separated from history merely by the passage of time, but temporary captives flowing with it, locked into the past by the unbroken chain of human events.

# A LONG HISTORY TO SORT OUT

The stream of human history in and around Grand Canyon stretches back at least 13,000 years and probably longer. Its significance is reflected in the archaeological record and in the oral histories of modern American Indians. The Hopi consider the canyon and all it contains to be sacred ground. For them *'ongtupka* is the ancestral home to 23 clans that have survived to the present. Other clans at Hopi have other points of origin. The Hopi call Puebloan archaeological sites *kiituvoyla*, house symbols. The remnants of these old Puebloan villages comprise some of the footprints of Hopi clan migrations. Today these footprints serve as cultural markers and remind the Hopi people of their obligations to the land. The trails they established to maintain contact with the canyon hold spiritual meaning, linking modern villages to ancestral homelands.

The Hualapai People have inhabited the south side of the Colorado River from Desert View westward to Lake Mead since time immemorial, calling *Mada Widita* (Spencer Canyon) in the far west end their point of origin before moving upstream. Prior to Euro-American contact the Havasupai were the easternmost band of the Hualapai Confederation. During the federal wars of conquest in the nineteenth century the government thought they were relatives of the Hopi, ultimately designating them an independent tribe after 1890 when those wars in the West were essentially over. Such is history. The Havasupai still make their home in Cataract Canyon within the depths of the Grand Canyon system, and consider themselves its guardians. Their Ram Dancers are a well-known dramatic presence and are emblematic of their long

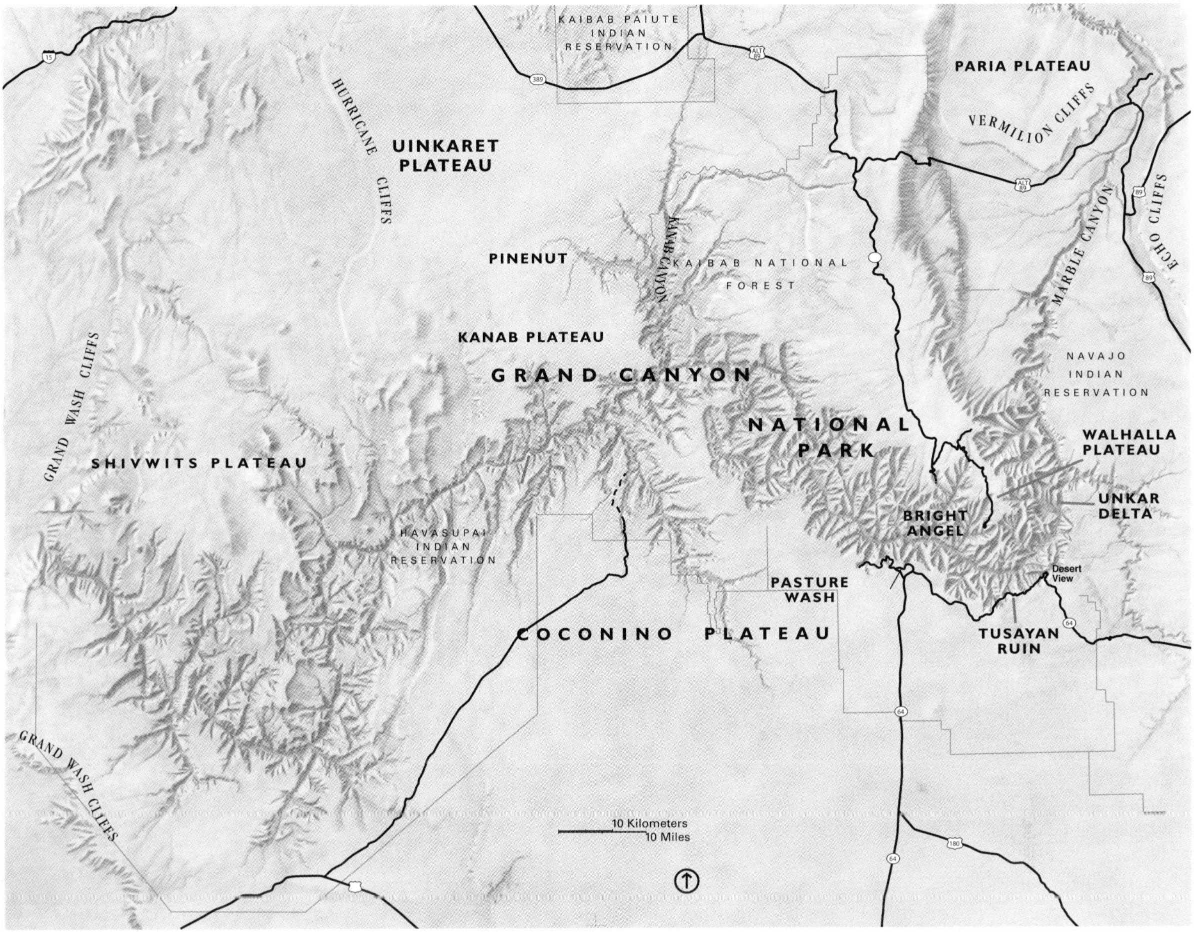


association with the inner canyon. The Yavapai, a sister tribe through language and common ancestry, are located south of Grand Canyon today; they also derive part of their heritage from roots in the west end of Grand Canyon.

The Southern Paiute inhabited the north side of the Colorado River for the entire length of the Grand Canyon from the time of the Big Drought (1180–1250) until the late 1800s. For them it was, and is, *puaxant tuvip*, holy land. Red clan symbols placed on rocks and old camp sites indicate their passing.

The place of Zuni emergence, *chimik' yana 'kya deya*, is located in the depths of Bright Angel Canyon in the vicinity of Ribbon Falls on the north side of the Colorado River. At some point the people living there crossed the river and left the canyon. They moved south to the base of the San Francisco Peaks, possibly Wupatki, where they lived for a hundred years or so before

Some details of Puebloan life have changed little over the centuries.

moving on to their current home to become the Zuni, all the while maintaining contact with the canyon and their Hopi allies. The Zuni refer to themselves as the *A:shiwi*, and they call the canyon *ku 'nin a' l' akkwe' a*.

The Navajo have been in the region for centuries, but only came into the eastern reaches of the canyon in the last 600 years or so. The *Dil zhe'e'*, a Western Apache group commonly known as Tonto Apaches, were also allied with the Hopi through trade. They ranged as far north as the South Rim of Grand Canyon to trade, raid, and collect. They appropriately called the expanse *ge da'cho*, edge of the big cliff.

Spanish conquistadors from Coronado's expedition, guided by Hopi men, peered over the South Rim in the late summer of 1540. A Jesuit missionary named Brother Kino visited the Havasupai in 1776, but before 1869 the canyon was the exclusive domain of American Indians. The territory, which the United States

Archaeologists excavate a site in the river corridor.

Bright Angel Pueblo lies less than fifty yards from the river bank, visible at the top of the photo. The pueblo has living areas, a storage structure, and a kiva. In cliffs behind the site, granaries held surplus corn and other dried goods. The room to the upper left of the kiva represents the first occupation of the site, A.D. 1050. The others were built after 1100 and used until about 1140.

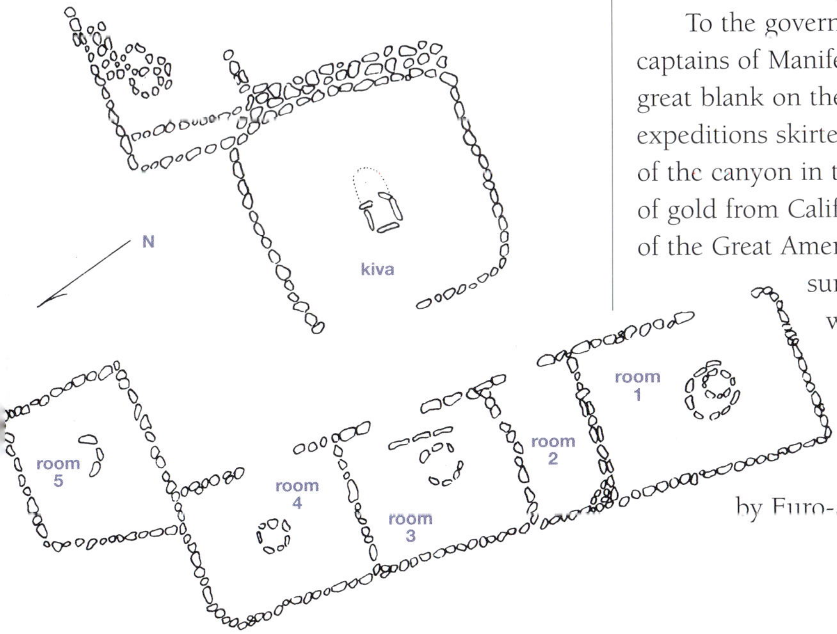


had obtained from Spain in 1848, was virtually unexplored by European-Americans. The Treaty of Guadalupe Hidalgo ceded vast stretches of the Southwest to the United States, including Arizona and the Grand Canyon region. This was done without the foreknowledge or consent of the tens of thousands of Tribal People spread across what would become eight states.

To the government in Washington and the captains of Manifest Destiny, Grand Canyon was a great blank on the map. Several U.S. Army expeditions skirted the 35th parallel, brushing south of the canyon in the 1850s and hauling wagonloads of gold from California eastward while the wide void of the Great American Desert was funneling railroad surveyors, geologists, and dreamers westward. Even with Major Powell's epic trip down the entire Colorado River in 1869, northern Arizona remained a land of pass-through and saw no permanent settlement by Euro-Americans until a full decade later.

*'ongtupka*
Hopi

*gik' mi giv tdé*
Hualapai

*puaxant tuvip*
Southern Paiute

*ku'nin a' l' akkwe' a*
Zuni

*bits' iis nineezi*
Navajo

*ge da'cho*
Western Apache

*Grand Canyon*
Anglo / European

Ribbon Falls on the north side of the Colorado River.

# AN ARCHAEOLOGICAL PERSPECTIVE

Because of its remote setting, professional archaeological work came to the canyon relatively late. The first serious project was a survey of the Walhalla Glades on the North Rim by Edward T. Hall Jr. in the 1930s. It was not until the 1960s that the foundation of our current archaeological understanding was laid, when Douglas Schwartz and his team conducted excavations on the Walhalla Plateau, in the river corridor at Unkar Delta (1967–1968) and in the Granite Gorge at Bright Angel Pueblo (in 1969), plus three decades of seasonal work throughout the canyon by Robert Euler. Numerous projects initiated by the National Park Service (NPS) in partnership with the Museum of Northern Arizona (MNA) and/or Northern Arizona University (NAU), including major excavations at Furnace Flats, have been completed since 1990.

In 1990–1991 as part of the belated Environmental Impact Statement for Glen Canyon Dam, an archaeological survey of the river corridor funded by the Bureau of Reclamation was conducted from the base of the dam to Lake Mead. The intensive survey documented more than 475 archaeological sites, tying together the excellent, but sporadic earlier work. The river corridor project, which worked within the historic high water zone stretching over 270 miles downstream, well illustrates the often-overlooked perspective that the canyon is not just a deep hole, it is also a very long one. The field work was accomplished by oar-powered inflatable boats over 10 months through the winter. There were four crews of three archaeologists, each with extensive field experience and specific areas of expertise, along with one boatman (or boatwoman) per crew plus a cook

A cache of Puebloan pots as seen when first found.

The restored cache of the Puebloan pots still holds pieces of the puzzle for archaeologists.

and their supply boat. The crews were headed by Helen Fairley, Pete Bungart, Terry Samples, and me, Chris Coder. Each trip typically lasted 20 days. Both sides of the river were scoured on foot. By 1994 a detailed report was produced, but it had a limited governmental circulation.

A five-year archaeological survey of Grand Canyon from 2003 to 2007, funded by the National Park Service's Service Wide Archaeological Inventory Program, generated a huge volume of information that required organization and analysis. This was accomplished by producing a multidisciplinary report that synthesized the data into something useful. This document, in the words of the primary author, Kim Smiley of NAU, will contribute to a much "higher resolution picture … of human antiquity" at Grand Canyon National Park, and that is the name of the game.

Between 2006 and 2009 this *higher resolution* was further enhanced by two projects: a joint (NPS/MNA) project excavating nine sites in the river corridor and a Bureau of Reclamation project documenting impacts to the river corridor by Glen Canyon Dam. The materials unearthed during excavation and the story they tell is amazing.

Until recently the archaeological focus has been on the Puebloan farmers, popularly and formerly known as the Anasazi, and their Basketmaker ancestors. They left behind the most durable stuff in concentrations; hence they have the biggest remaining physical signature. Because of that signature John Wesley Powell observed one of their pueblos at the mouth of Bright Angel Creek during his voyage down the Colorado River in the summer of 1869. Major Powell noted in his journal that he had encountered *"the ruins*

**STONE TOOLS**

Stone tools are divided into distinct categories: chipped and ground. Chipped stone tools include spear, dart, and arrow points; knife blades; punches; scrapers; awls; gravers; and a variety of other items. They are generally produced from fine-grained or crystalline rock that can be obtained from within the canyon. The Kaibab and Redwall Limestones contain seemingly limitless quantities of workable chert. The cobble bars along the river and benches below the rims hold an enormous supply of knappable rock. Highly valued obsidian and workable rhyolite are common to the immediate region with several sources as close as Flagstaff and Seligman, Arizona.

The metate and mano are used to grind grains into flour. Such groundstone tools are made by pecking and grinding pieces of stone into the desired shape.

*of two or three old houses, which were originally of stone laid in mortar. Only the foundations are left, but irregular blocks, of which the houses were constructed, lie scattered about. In one room I find an old mealing stone, deeply worn, as if it had been much used. A great deal of pottery is strewn around, and the old trails, which in some places are deeply worn into the rocks, are seen."*

We have come a long way since Major Powell's fleeting observations were made. Today our understanding of Puebloan life has been fine-tuned by the bright light of modern science, the traditional knowledge of tribal elders, and dozens of dedicated professionals.

Archaeology is defined as the scientific study of past cultures, their material goods (artifacts), and associated human behaviors. Linking a people's everyday possessions to the greater complexity of their culture is tricky. It can be like trying to get an accurate understanding of the lives of our great great grandparents by deciphering a jar of old buttons, a tattered quilt, an axe handle, and a kerosene lamp. The objects tell us something, even something important, but what?

The archaeologist is a cultural detective who must draw upon several disciplines to reconstruct a picture of the past that is as complete as possible. The past becomes clearer with an understanding of landforms (geomorphology), soils, climate, formal excavation, and the laboratory work it generates, including radiocarbon (C-14) and tree-ring dating, the study of pollen, phytoliths, seeds, and fibers, the microscopic wear patterns on tools, and ceramic and genetic analysis. We are always looking for evidence of change and patterns, and the patterns those patterns create.

## ARCHAEOLOGICAL DIVISIONS

Archaeologists in the early twentieth century divided the prehistoric Pueblo into temporal divisions based on advances in architecture, land-use patterns, point typologies, and design styles found on pottery. These divisions serve as a framework for studying the past, but would be meaningless if presented to a farmer in Grand Canyon a thousand years ago. The designations of Basketmaker II, III and Pueblo I, II, III relegate an entire culture to stylistic changes in material goods and do not address the more difficult task of understanding economics, migration, social fabric, and religion.

The traditional cultures of Grand Canyon were elegant extensions of the rocky landscape they inhabited. These cultures left behind evidence that is often subtle and not easily interpreted. Because they lived in direct contact with the earth, through human ingenuity and toil, they developed the skills required to transform the raw materials at hand into the routine of their daily lives. Stone tools, cordage, ceramics, plant medicines, food gathered and cultivated, habitations and storage, rabbit nets and fur blankets, turkey bone whistles, serpentine pipes and so on are examples of their creativity and resourcefulness.

Time and the elements are pervasive and brutal, stripping away the flesh and soft parts without bias. Perishable materials—leather, wood, sinew, plant fiber (baskets), flesh, and hair—fade quickly with the passage of time. In arid inner-canyon caves and rock shelters, such artifacts are occasionally preserved, but that is the exception and not the rule. Often we must try to decipher a vision of the past based on an assemblage of artifacts that represents the survival of the hardest items—pottery sherds, projectile points, and stone hand tools—but not necessarily the most informative. Petroglyphs and

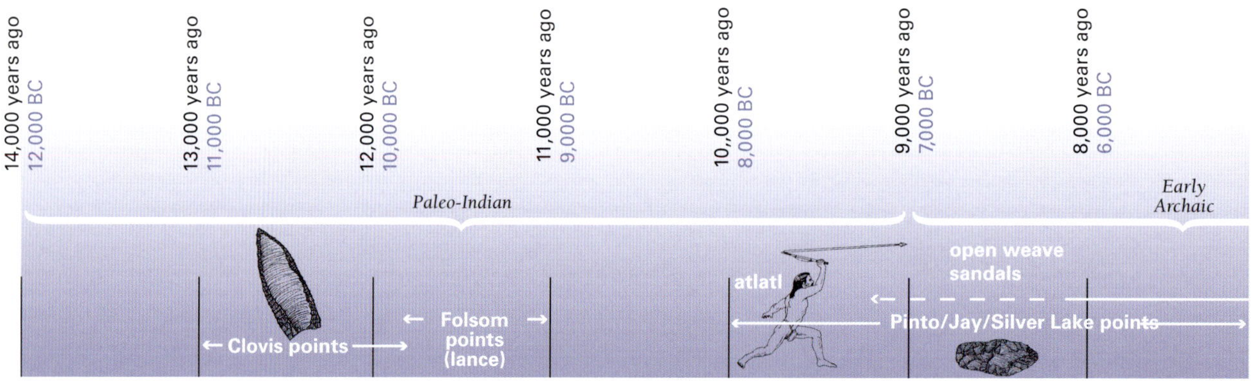

pictographs are additional clues, but the intriguing symbols and abstract images often create more mystery. Current ethnographic work with tribal experts adds more critical detail to the story. Today archaeologists rely on American Indian expertise for insight into what academic culture calls prehistory and what tribal people simply consider their history.

Geologists use a concept called facies change to explain the way rock formations undulate over time across the surface—thick in one place, thinning in another direction, ultimately disappearing altogether as the deposits become more distant from their source materials, eventually merging with other formations emanating from other places. To envision this requires a three-dimensional perspective. The same is true of cultures. They can be hard to trace as they expand and contract, merging with other cultures over time. Where did they start out? What do they have in common? What is unique? How do we sort out the difference? With time as a factor, it becomes a four-dimensional exercise.

In the ongoing tradition of Western Civilization, our scholars divvy up the past into segments, each one of which can be defined by characteristics that set it apart from others: the Paleolithic, the Archaic, the Bronze Age, the Iron Age, a big jump to the Age of Reason, another big jump to the Roaring Twenties, and then the Great Depression when all the cars were black. We carve time into centuries, decades, months, days, and so on as a cupboard for storing information in a logical fashion. It is simply a chronological filing system.

In the American Southwest the Paleolithic period is differentiated from Early Archaic sites to a degree, based on observable changes in projectile-point technologies. The Early Archaic is distinguished from the Middle Archaic partly based on differing methods of sandal construction and changes in regional climate. The Late Archaic is discerned once again by point types, new rock art styles, the beginnings of agriculture, and at Grand Canyon in particular the appearance of distinct split-twig figurines. These are artificial lines we have imposed on the

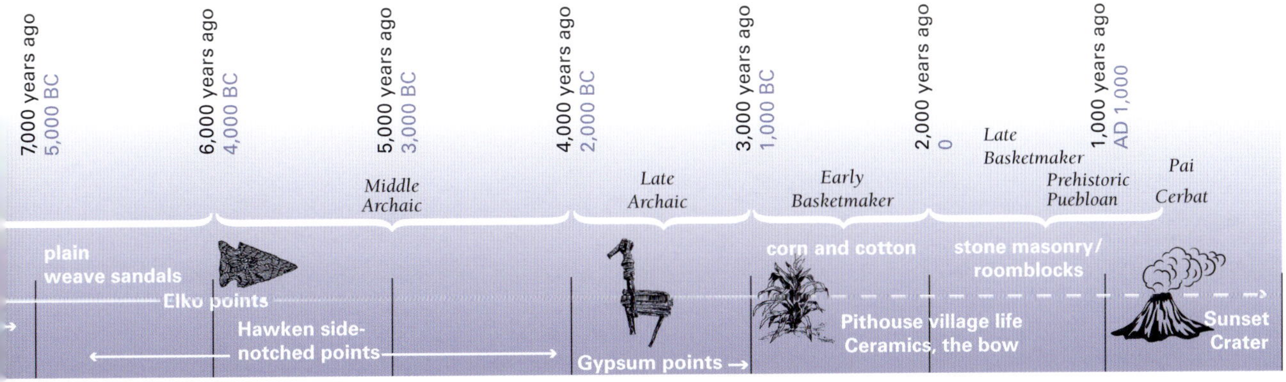

past with indifference to the actual people who lived through them.

Puebloan sites can be distinguished from Late Archaic sites by lifestyle changes—surface structures, the presence of ceramics, and the suite of artifacts associated with greater dependence on maize (corn). Culturally, events develop a momentum of their own, more complex than can be figured out precisely, with one event influencing the many, and the many the one. As is true of our own lives and the lives of our families, the farther we go back in time the more hazy events become. With all that in mind, we can begin to look at the march of human history in Grand Canyon.

## Paleo-Indian Hunters

From about 13,000 years ago back to nearly 2 million years ago, so much of the earth's water was tied up in glacial ice that sea levels were periodically as much as 400 feet lower than today. Long-term cycles of glacial advances and retreats typical during this time regularly exposed a called the Bering Land Bridge and known today as Beringia, which physically connected Siberia and Alaska over a 900-mile-wide expanse. As the glaciers grew and water was turned to ice, sea levels dropped and the previously submerged continental shelf slowly appeared—as the glaciers melted and the sea rose, the shelf disappeared under the waves. This occurred several times between 40,000 and 13,000 years ago, allowing hunting groups to periodically migrate from Asia into interior and southeastern coastal Alaska.

By 13,100 years ago the tools and rare camps of these hunting groups appeared all over the Intermountain West and the Great Plains. Known as the Clovis people, they were named after the town in New Mexico where their meticulously crafted spear points were first discovered in the 1920s (before C-14 dating) with the bones of now-extinct mammals. Professionals studying early humans in the Americas now believe that Clovis People entered the lower 48 states by dropping down the Pacific Coast from Southeast Alaska, fish hooking around Baja California, entering the Colorado River Delta, following that river system upstream, taking the Gila fork into New Mexico, and ultimately spilling out onto the Southern Plains in Texas and Oklahoma, where they created the Clovis blade technology that defines them as a culture. Previously it was assumed they had brought the technology with them from the far north through the ice-free corridor in Canada,

but modern scientific techniques and a raft of new discoveries have laid that old idea to rest.

Scientific evidence of the peopling of the new world by mechanisms other than the Bering Land Bridge is absent. Sites excavated at the tip of Chile by archaeologist Tom Dillehay have extended New World human history back more than 17,000 years. It is broadly accepted that humans have been in the New World much longer than previously recognized—in small numbers, perhaps as long ago as 26,000 years in central Alaska. Polynesian migration by sea is a given, but that happened much later across the Central and South Pacific, after 500 A.D. They may have crashed into South America on more than one occasion, but Polynesian culture did not take in the New World.

The Clovis and subsequent Folsom cultures were sophisticated big-game-hunting people as well as skilled foragers. Evidence of their success and passing appears throughout the United States. The Colorado River Basin contains evidence aplenty of the Paleo-hunters. Camps have been found along the San Juan and Green Rivers in Utah, as well as the rocky benches of the Little Colorado River, but at Grand Canyon the traces are confined to a few spear points, which may have been curated by mobile hunters from other places and brought to the canyon as a part of their own tool kit. They were here, but passing through in such small numbers, that what little they left behind has been ground to dust, covered over, swept away by flood, or scavenged and re-worked by those who came along later.

We know that Paleo-Indian people were few in number, a small group here, a band of families there. They lived life mostly on the go, moving without a map across the landscape, searching for or following big game. Their possessions were mainly perishable and renewable: leather, bone, wood, and plant fibers. With a mobile hunting people, what is not essential gets left behind and the centuries meticulously swallow it up.

It is not known exactly how many people comprised a working band or unit. It no doubt varied—perhaps more than a dozen and fewer than forty, depending on the available resources—and those preferences are not preserved in the record. Humans are cultural beings and cultures do not exist in a vacuum. Each band was an independent cog in a greater society. These extended-family bands would connect with other related groups at specific times and places to trade goods and stories, reinforce old connections, and arrange relationships that transcended the seasons.

They were walking the tightrope of changing times. The world was slowly warming up. The glaciers were steadily melting. Analysis of Antarctic ice cores and deep ocean sediments indicates a radical change in the global climate right at 12,900 years ago. An episode of intense melting stopped on a dime (less than a decade) and then the Ice Age came back with a vengeance for almost a thousand years. This mini ice age is known scientifically as the Younger Dryas and was until recently an enigma with no obvious cause. It is difficult to explain, but is being worked on assiduously.

After the Younger Dryas the landscape went back to rearranging itself accordingly. It is from this time that most of the cultures around the

## END OF PLEISTOCENE

It is accepted science that Ice Ages come and go based on the predictable long term cyclic wobble of the earth, which determines how much annual sunlight hits the northern or southern hemisphere. Less sunlight and the glaciers advance, more and they melt back. What is harder to explain is why the gradual warming and retreat of the glaciers occurring from 20,000 to 13,000 years ago experienced an instantaneous and radical melt-off right at 12,900 years ago, followed directly after, without missing a beat, by a severe, but short lived resurgent 1,200-year Ice Age known as the Younger Dryas. The geological and archaeological records are quite clear on these events. The sudden melt-off at 12,900 years ago truncated the flourishing Clovis Culture in the blink of an eye, and the rapid return of a colder climate doomed the already over hunted and faltering Pleistocene megafauna; mammoths, mastodons, giant sloths, cave bears, large wolves, Bison antiquus, glyptodonts and scores of smaller species, who had survived the previous two million years of extreme glacial and interglacial swings. What precipitated this radical melting followed by the Younger Dryas mini–Ice Age has been a scientific mystery for years, but now there is a likely explanation.

Recent scanning electron microscope studies conducted on materials collected from terminal age Clovis sites revealed microscopic diamond sherds melted into the surfaces of the tools and bone. These micro diamonds can only have had a cosmic source, which suggests a large meteor struck the retreating ice sheet in the northern hemisphere, possibly at Hudson Bay, causing catastrophic melting worldwide, simultaneously blocking out sunlight and creating atmospheric conditions mimicking several centuries of a nuclear winter manifested as a return to a colder climate, the Younger Dryas. Barring a better explanation this concept is the most plausible mechanism for the catastrophic melt-off and subsequent return of the ice. The unexpected conditions of the Younger Dryas set indigenous cultures and the Pleistocene megafauna back on their heels, but in North America south of the shock zone, people survived, regrouped, retooled, adjusted to the new normal, and emerged in the western United States as the Folsom Culture. As the Younger Dryas ice disappeared and the climate began to warm back up 11,700 years ago plant and animal species in large numbers had disappeared with it, but humanity, albeit morphed, survived and carried on.

Mammoths, along with giant sloths, short-faced bears, and wolves were hunted by the Paleo-Indians of Grand Canyon country.

The Colorado Plateau, separating the high country on the north and east from the deserts of the Basin and Range to the west and south, was home to generations of Native American people.

world enshrined the occurrence of the great flood(s) into their cultural histories and legends. Such was the severity and scale of the change that the atavistic memory of it remains widespread. The subsequent floods of various magnitudes in the annual cycle of human experience would just reinforce the effect of the original monster in people's minds. Think of the catastrophic floods caused in South Asia (Phuket, Thailand et al.) and more recently in northern Japan by tsunamis. Because of where we like to live, floods are simply a cost of human business. During episodes of intense melting as the Ice Age geared down, the Colorado River would have been raging through the canyon for decades at a stretch, not fordable in any season.

Pleistocene megafauna—mammoths, giant sloths, camels, North American horses, short-faced bears, and giant wolves—were slowly passing away with the glaciers. The hunters' success at bringing down the largest game helped seal their own fate by hastening the decline of the mammoth herds. As the Pleistocene wound down, so did the size of the spear points; small game did not require large projectile points. Late Paleo-Indian Folsom points were half the length of the elegant Clovis Tradition blades. The older order was in transition; between 12,900 and 10,000 years ago each successive generation was experiencing the downward spiral of diminishing returns with the shrinking size of available game.

Groups of hunters living on the Colorado Plateau changed along with their world. The transition was not as difficult as it would first seem. Paleo-Indian people had been skimming the big calories, but there was still plenty to eat out there. They fine-tuned their hunting strategies by downsizing their Big Critter getters into Little Critter getters to acquire deer, bighorn sheep, and smaller, quicker animals. Hunting deer might not have been as dangerous as taking down mammoths, but it would by caloric necessity have to occur more often with different technologies and tactics. To replace a mammoth it could take, depending on the size of the deer, between 50 and 100, or more than 2,000 fat rabbits. It also took new and variable styles of cooperation, changing the dynamic in the community and to some practical degree the ties that bound groups of families together. Those adaptations, however small, in a working band are transferred directly to the hearth and bowl.

Changes in tactics were manifested by changes in the tool kit and those adjustments were in turn reflected in the archaeological record by new patterns. In the age that followed, the hand-held spear hafted with large fluted blades was replaced by the atlatl (a spear-thrower) and shorter, lighter darts tipped with smaller projectile points. Plants became a more critical source of food. The ground-stone hand tools and grinding surfaces required to process seeds, nuts, and other vegetal materials appear

This broken Folsom blade from Grand Canyon is tantalizing evidence that Paleo-Indian hunters were moving through the Nankoweap Basin more than 10,000 years ago.

**DIMINISHING RETURNS**

Because projectile points possess attributes tuned to specific tasks, they reflect the cultures that produce them. To bring down large Ice Age game, Paleo-Indian people fashioned large blades hafted to spear shafts. Archaic people designed dart points for deer, bighorn sheep, and other medium-sized game. Late Basketmaker expanding stem points were scaled down to accommodate the new bow-and-arrow technology. Prehistoric Puebloan points were downsized further, and the Hualapai (Cerbat) and Southern Paiute people who came along later manufactured even smaller points, reflecting the world they lived in.

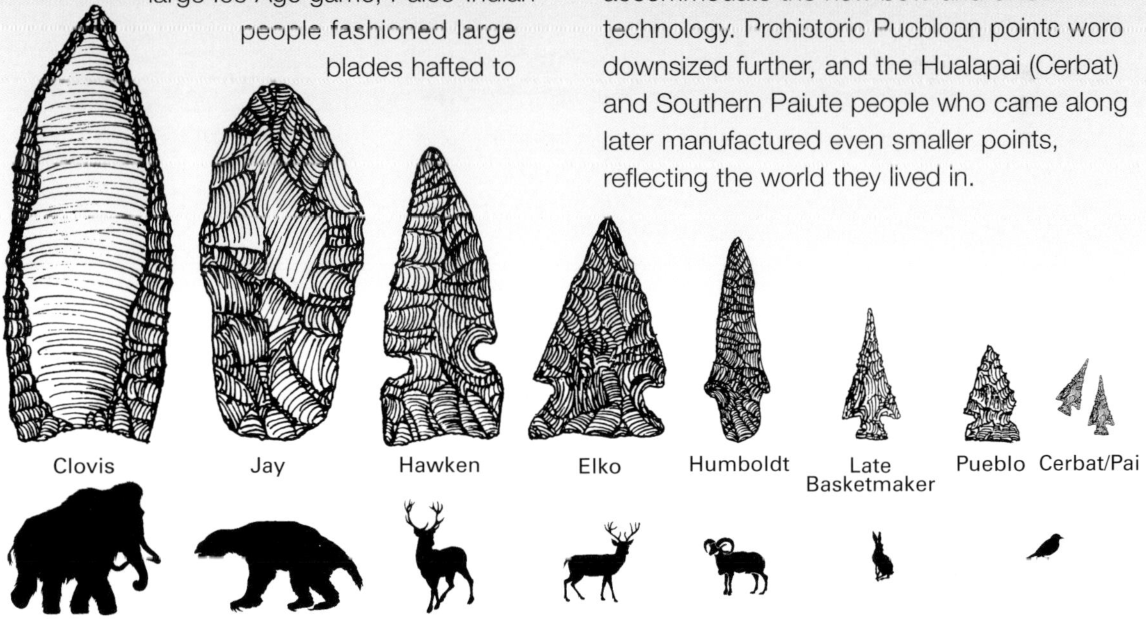

more often in the record after the demise of the large Ice Age game. The patterns of regional movement changed and in the American Southwest the mobile hunters began to establish home territories.

Folsom, Humboldt, and Jay style blades and projectile points belonging to Late Paleo-Indian stone tool traditions are found across the uplands of Grand Canyon National Park. This indicates that small groups of people were filtering into or through the region even as the big game died out. Their light hand on the landscape combined with their small population left only a tantalizing portal into their passing. The Paleolithic People didn't get beamed up; they didn't get fire and secret knowledge from aliens. They were humans. Humans are smart (generally)—some stayed and adapted in place, some went somewhere else and did the same. Traveling, adapting, and surviving remains our prerogative and legacy as humans.

## The Archaic Period

By 9,000 years ago, more people had entered the Grand Canyon region from the Basin and Range Province to the northwest with all the trappings of Archaic culture: atlatl and darts, open-weave sandals, seasonal habitations, ground-stone tools. Indications of human settlement in Grand Canyon country during the long centuries of the Archaic are extensive. The Archaic period in the American Southwest covers such an expanse of human history that it has been divided into three parts: Early, Middle, and Late. These broad divisions are based on several factors: changes in projectile point technology, alterations in climate, and regional shifts in population.

Early Archaic culture is transitional from Paleo-culture, reflecting the loss of the large Pleistocene game animals and a warmer, drier climate, relying more on smaller game and depending on stored plant calories in the form of nuts, seeds, and tubers to get through the winter. In spite of these major inconveniences, the human population on the Colorado Plateau increased during this period. People slowed down a notch. The pace of life was more conducive to staying in the same camps longer and hence preserving more of what the human experience produced in that place. So the record from these times is more complete and a little less mysterious than the Paleo-Indian. About 7,000 years ago the climate became drier still, signaling the beginning of the Middle Archaic drought, known as the Altithermal, which would last off and on for the next 2,000 years.

As the West dried up, erosion accelerated. With each generation there was less plant cover holding the softer sediment in place. Sand was released into the drainage systems in large quantities. The primary rivers became sluggish and smaller streams clogged up. The reliable springs turned into seeps and the seeps disappeared completely. Gullies dissected the dry landscape. Old terrace systems were etched back and removed. The wind gathered sand into dunes and choked every empty space. It might have been possible in many years to wade across the Colorado River in Grand Canyon in specific places during many months of the year.

Perceptible change to the landscape would have occurred over the period of a person's lifetime. Over three lifetimes it changed dramatically. People would have to go five times as far to get half as much water, game, or seeds. At some point people say enough is enough. One option would be to leave the region and take their chances on the road, which is risky business, but it might also mean a better life. Just think of the exodus from the American Dust Bowl in the 1930s on a smaller scale, but without any old trucks or hamburger joints.

Another strategy would be to tough it out in place. For those choosing to stay, the large rock shelters were a poor option and were mostly abandoned. As the springs associated with the shelters failed, people were forced to relocate to the dwindling sources of permanent water. The groups that had remained to weather it out with the landscape refocused their efforts on the shriveling resource base they were confronted with. Understanding the local and often radical changes in elevations and those connections to the seven biotic communities of the Grand Canyon, was key to their survival. The ability to glean a living from the landscape was based on the knowledge of plant and animal communities and in turn those communal relationships based their own survival on the same factors—elevation, water, and season.

So going up-slope was a pretty good tactic; more water might mean more forage, which might translate to more game and hopefully more protein. As an example the San Rafael Swell in central Utah served as a high-altitude refugium for people and game during the worst years of the Altithermal. The Grand Canyon with its classic staircase ecology would have served the same purpose. Population studies from across the Intermountain West for this time frame are mimicked at Grand Canyon. The few people still moving between the canyons and plateaus would occasionally camp on the soft sand sifting into caves that had been permanently inhabited throughout the Early Archaic. These environmental refugees would build fires, repair their goods, rest, and move on, leaving traces of their passing on the surface, some of which would be incorporated into the stratigraphic record.

The highly erosive nature of the Grand Canyon system constantly rearranges the surface and reworks the sedimentary deposits. Soils older than 5,000 years are rare. Nevertheless, Middle Archaic sites have been recorded in a tributary of Glen Canyon. Bowen's Canyon, located upstream

The introduction of the atlatl (a spear-thrower) greatly improved the odds for the Archaic hunter.

An artist's view of the Archaic people's hunting-and-gathering lifeway.

A large cache of split twig figurines were found in Stanton's Cave

from Lees Ferry, harbored a community that endured the hard times by taking up residence in a place with a permanent source of good water. Significantly in 2008 the GRCA/MNA River Corridor Project excavated deeply buried cultural deposits at the Axe Handle site below Lees Ferry, as well as at Arroyo Grande, Tanner Delta, and below Unkar, all dating to the Middle Archaic. These sites are similarly situated to make best use of the permanent water of the Colorado River.

By 4,500 years ago the severe dry times were waning and populations were growing. A good deal of cultural evidence from the Late Archaic can be found at Grand Canyon. The Gypsum style dart points and blades these people used are commonly found in the Park north of the river.

One of the best visual examples of Late Archaic culture is the elaborate pictograph panel known as Shamans Gallery located on the North Rim.

Perched 50 feet above the Colorado River in a narrow stretch of Marble Canyon is a large opening in the Redwall Limestone named Stanton's Cave, which reveals a story of Late Archaic hunters who came and went in this remote spot more than 4,000 years ago. Excavations in the cave uncovered dozens of split-willow-branch figurines; some were run through by small twigs representing dart shafts.

These shadows of human behavior are found across the Great Basin into Nevada and south of the canyon in the dry mountains above Clarkdale, Arizona on the Verde River, telling us that these particular people were part of a larger ranging culture and that families or clans moved around.

The Late Archaic people of Grand Canyon acquired life's necessities from the stacked resources between the river and rim country. They hunted bighorn sheep in the gorge, scoured side canyons for mineral pigments and medicinal plants, harvested pinyon nuts, and hunted deer in the uplands. They had dogs, traded for goods outside the region, produced and wore buckskin clothes, and made personal adornment. They did not make or use pottery. They made fabulous baskets. They were experimenting more than 4,000 years ago with corn and squash, which had come north from Mexico on the Uto-Aztecan trade circuit. They pondered their relationship with the cosmos, as evidenced in their complex pictograph panels. Their skeletons are very rare. Not much else about them is known. Like the Paleo-hunters before them, their goods were mostly perishable. So we are—again—faced with defining an entire

Figurines, made by shaping and wrapping two halves of a split willow twig into the likeness of a deer or bighorn sheep, are found across Nevada, parts of Utah, and in Grand Canyon north of the Colorado River. Strikingly similar to images seen in Archaic petroglyphs, they are diagnostic of Late Archaic culture. Usually found in dry caves, the figurines are of varying dimensions, but most are palm-sized. They were charms used in pre-hunting rituals.

## PETROGLYPHS AND PICTOGRAPHS

Rock art refers to the various designs and symbols placed on a rock surface by the human hand. Some individuals take exception to the term. The form is artistic, but to the Native American calling it art is misleading at best and at worst offensive.

Two major categories are recognized: (1) Petroglyphs are pecked or scratched into the rock face creating a negative surface and revealing the lighter rock underneath; (2) Pictographs are created by applying pigment to the rock, leaving a positive color image on the surface. The most common colors are black (charcoal), red (hematite), white (kaolinite), yellow (limonite), blue and green (oxides of copper).

Rock inscriptions are not language, and the meanings are as diverse as human behavior. Some are the enigmatic visions of a shaman emerging from an extended trance, some are clan symbols signifying use or a claim, some are stories, some are directions, and others that appear to be only smudges or colored streaks are the prayers of individual men and women.

Glyphs and graphs act as cultural and individual signatures, adding color and flesh to the archaeological skeleton of old walls, sherds, and chipped stone. On occasion styles can be connected to a specific language group or cultural tradition and with some success can be dated. Hopi clan symbols are present at the Salt Cave in the River corridor, and an unusual petroglyph placed on a bedrock bench below Lees Ferry is associated with the *Newe:Kwe* (Galaxy Fraternity) of the Zuni. A pictograph panel in the west end of the canyon commemorates the Paiute Ghost Dance of the late 1800s. Old Puebloan and Late Archaic panels are found throughout the canyon region.

Detail of Shamans Gallery pictograph.

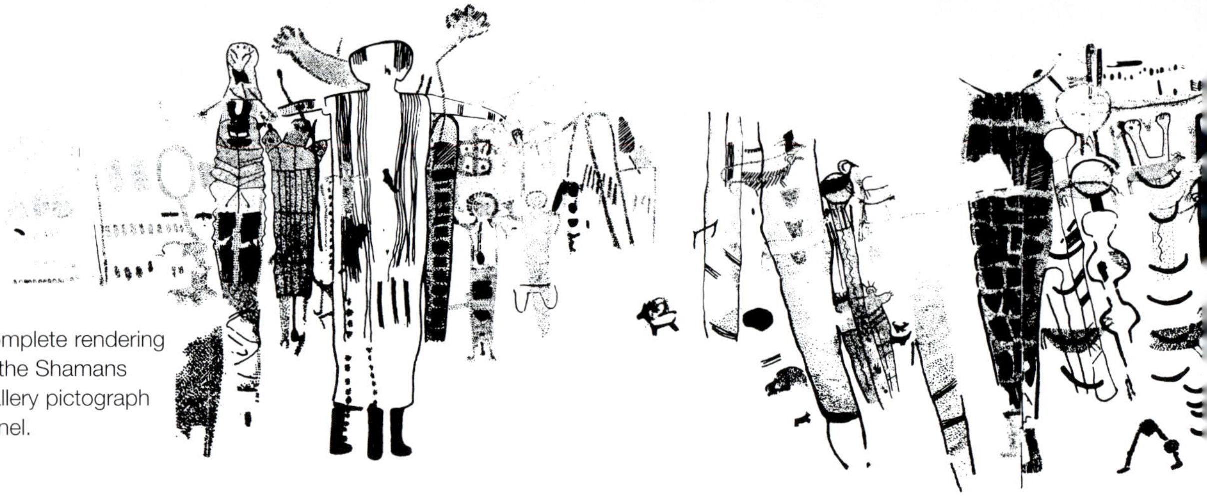

Complete rendering of the Shamans Gallery pictograph panel.

Fremont pictograph

Puebloan petroglyph

Puebloan petroglyph

Hopi clan symbols

Puebloan petroglyph

***The preservation and protection of these prehistoric records is a vital part of the mission of the National Park Service. Their cultural significance to tribes and the archaeological record is still being unraveled.***

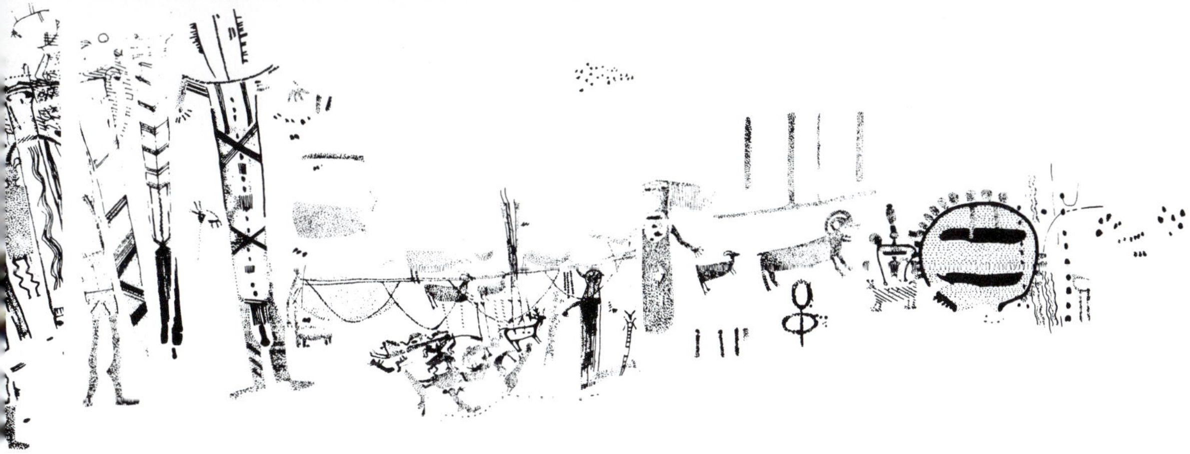

people by a few tools, some figurines, and an occasional thought-provoking pictograph panel.

They lived quietly on the landscape they were intimate with, not without strife, but without a lot of static either—no tax forms, no mechanical clocks, no canned food, no traffic. If they got a compound fracture or appendicitis or a woman had a breech birth they were in big trouble. They possessed an understanding of plants and animals, water and rock, climate and season that is really beyond our grasp. Theirs was not a scientific perspective, but one of intuition and experience accumulated by trial and error for literally millennia. Knowledge was passed generation-to-generation by imitation and the spoken word when life, landscape, religion, education, and culture were synonymous.

## The Basketmakers

The earliest corn-growing people at Grand Canyon, commonly known as the Basketmaker culture, correspond to the very end of the Archaic time frame. They cultivated corn and dabbled in squash of many varieties, but game and wild plant foods still dominated their diet. These people were scattered around Grand Canyon in family camps and seasonally occupied villages. Botanical evidence from samples collected at Nankoweap Delta and above Tanner Rapid indicates that they were growing a variety of native cotton in the river corridor. They lived in rock shelters where available, otherwise in expedient shelters and pithouses, which were underground homes that were entered through a hole in the roof. Pithouses were well insulated, being cool in the summer and warm in the winter.

### COMMUNAL USE OF RABBIT NETS

The weaving and use of a rabbit net were a communal affair requiring cooperation and coordination by the many for the benefit of all. The net was taken care of by an individual or family, but it was 'owned' by the community and its proper deployment and successful operation was a team effort. These brilliant devices were intricately made and aesthetically beautiful. The one pictured here is of Late Basketmaker design and contains more than a mile of hand-twined cordage. Rabbit nets were still being used by Hualapai and Southern Paiute people into the early 20th century.

Basketmaker II rabbit-net from White Dog Cave, Arizona, before unwrapping.

**CORN: A CHANGE IN THE WIND**

During the Late Archaic the opportunity for groups of hunters to scale down the degree of mobility in their lives was at hand. Corn (maize), beans, and squash had already been under cultivation in the Valley of Mexico for at least 4,000 years. But it was not until 4,500 years ago that people moving northward out of the Mexican highlands brought corn into the American Southwest. By 4,000 years ago it was being grown locally on the Colorado Plateau south of the Colorado River. By 2,500 years ago it had jumped the Grand Canyon onto the Arizona Strip and northward into Utah. Corn changed everything.

Alterations to a pattern of life established over centuries, like stopping to plant corn, are not made casually, and it is not something that happens overnight. The repercussions are numerous. Nutrition, social fabric, spiritual focus, and the routine of numerous daily tasks must change to accommodate the new order that comes with village life and dependence on maize. People must ease their way into it. For some these changes were not acceptable and not all of the groups living in northern Arizona chose the road that led to small-scale farming.

Historically, some Western Apache and Navajo families grew corn, others did not. The choice was based on personal preference, and that may be a clue to how certain people eased into this way of life at Grand Canyon.

Items that set the Basketmakers apart from other cultures were cradleboards with soft headrests, square-toed sandals, beautiful woven blankets and ponchos, subterranean slab-formed clay-lined storage cists, intricate baskets, and curved throwing sticks for hunting small game. They wove with such skill that their baskets were watertight. In fact, they cooked by adding heated rocks to concoctions of water and foodstuffs contained in such a vessel to make soup and porridge. Seeds and nuts were parched by flipping them in a wicker tray with small, hot stones.

They did not begin to make pottery until about 1,700 years ago. About that same time the Uto-Aztecan culture train brought bow-and-arrow technology south, quickly replacing the atlatl and dart. The bow was likely brought to the Southwest by Uto-Aztecan speaking Shoshone hunters from the Columbia Plateau in the American Northwest. This masterful innovation

An artist's view of a Basketmaker community with pithouses.

further downsized projectile points. The bow made the acquisition of game a less communal affair by making the individual hunter more effective. It also changed the tactics of raiding and local warfare. A true archer with a big jar of water and a family for motivation would have been difficult to dislodge from a good defensive spot. It took more than two centuries for this new technology to jump the Colorado River and move southward off the Colorado Plateau into the low deserts. Many centuries later, after 1514 A.D., far to the south the Spaniards encountered Aztecs and Incas still using massed warriors with atlatls and darts.

By the end of the ninth century A.D. most of the farmers on the Colorado Plateau had traded the pithouse for the above ground stone roomblock. In the centuries to come, some of the Basketmaker groups that would become known as the Anasazi (now Puebloans) retained the traditional pithouse design as the basis for the subterranean *kiva*.

# A PROFILE OF PUEBLOAN LIFE

By 750 A.D. what is today recognized as Basketmaker culture was all but replaced by the lifestyle of the pueblo. Like the evolution of the thirteen original English colonies into the Euro-American United States, it was a process, not an event. Remembering the facies change concept used by geologists, we can say that Basketmaker culture grades into Puebloan culture across the region.

The ancient Puebloans were not a homogenous people, although on the surface they appear to be. Archaeologists have differentiated them into several geographical traditions based on social organization, ceramic styles, trade associations, architecture, and other esoterica. The traditions are Kayenta, Mesa Verde, Chacoan, Virgin River, Little Colorado River, and Sinagua. This at first seems like hair splitting, but there are notable differences and archaeological detective skills are often required to sort it out.

Cultures, like the families who form them, have lots of overlap with each other, but it is often nuance and subtlety in the physical record that sets them apart. At Grand Canyon, Kayenta and Virgin Puebloan traditions blend and merge on the north side of the Colorado River, just as the Kayenta and Cohonina briefly touch in time and space on the south side. Most of the other western branches of Puebloan People probably spoke different tongues of a broader language family believed to have been Uto-Aztecan. A notable exception to this are the *A:Shiwi* (Zuni) whose language is its own family, unlike any other, further indicating the variety of cultures that came down from Paleolithic times with their own traditions and languages intact. We know

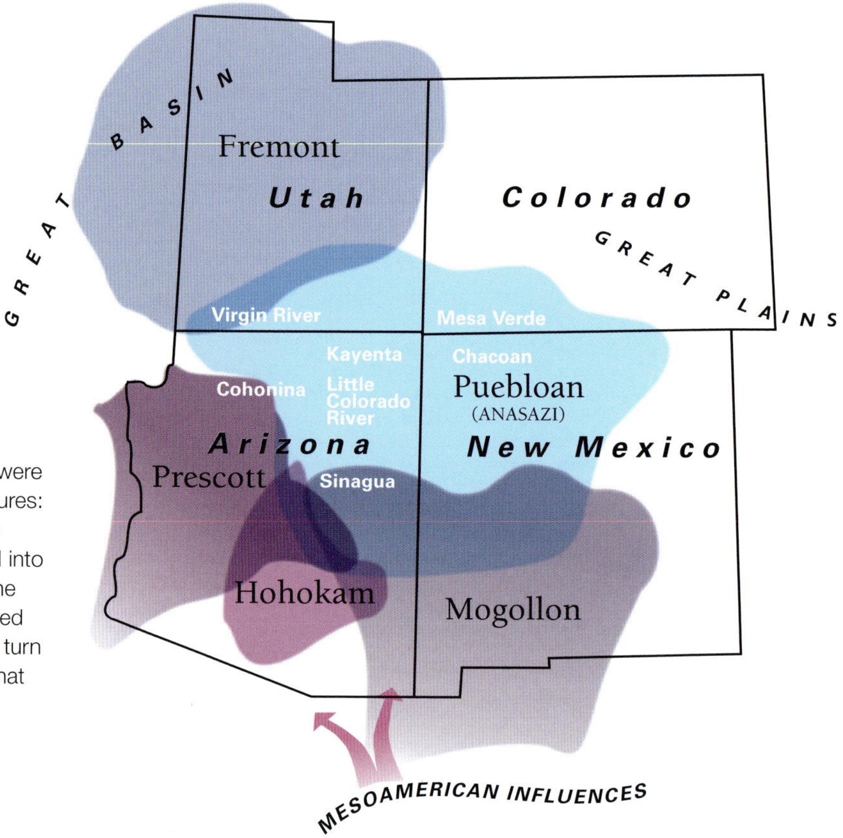


The Grand Canyon Puebloans, comprised of several traditions, were surrounded by other unique cultures: the Mogollon, the Hohokam, the Prescott, the Fremont northward into Utah, and the various tribes of the southern Great Plains. They traded and warred with these people in turn and placed names on the land that are today mostly forgotten.

that it was common historically in America for native people to speak multiple languages. This was assuredly the case a thousand years ago and farther back amongst trading people like the Puebloans and their neighbors.

These distinct groups were aligned by trade, the physical and social importance of maize, material culture (their stuff), and possibly religion. The relationships between and within Puebloan centers changed over time and oscillated between amicable alliances and interdependent trading to raiding and dominance through religious and economic control, the likes of which spread outward from and retreated to Chaco Canyon as that center rose to prominence and then fell, between 1030 and 1140 A.D.

Prior to a thousand years ago Puebloans lived in the uplands along the abrupt rim country and farmed in the river corridor, tending small plots of corn, squash, and cotton as conditions would allow. In 1991 a flash flood at Chuar Canyon uncovered a classic assemblage of pottery sherds in a cut-bank dating to the Pueblo I Period, nearly 1,200 years ago, revealing the presence of a prosperous farm family making a living along the river 200 years before the Puebloan surge and farming miracle at Grand Canyon (after 1040 A.D.)

Around a thousand years ago the climate began to shift once again, this time to the advantage of farmers. A slight increase in the amount of seasonal precipitation allowed corn, beans, squash, and cotton to be grown with reliability in more places. This change in the rain belt temporarily allowed Kayenta farmers to expand across the Colorado Plateau wherever a crop could be coaxed from the soil. This included the San Juan River country in Utah, the Kaibito Plateau, and downstream along the Colorado River through Glen Canyon into eastern Grand Canyon. Their burgeoning population also put them in direct line of site with the violent hegemony soon to emerge from Chaco Canyon.

Chaco Canyon, to the east in northwestern New Mexico, rose to dominance around 1,000 A.D. and began to exert their influence on a regional level at about the same time. As the Chacoan warrior/priest elite extended their corn tax outwards on their spoke-straight roads across the Plateau, some of the vulnerable Kayenta farmers may have opted to move farther west to the relative safety of the Grand Canyon, beyond the tax collectors' easy reach and the violence accompanying their mandate. The increasing

Thick-plaited yucca fiber tells us the sandal above is prehistoric Puebloan. Sandals, like the one at the right, of the older Basketmaker culture were more painstakingly made with cotton cordage.

*Anasazi* is the term used in the past to describe various maize-dependent Puebloan cultures inhabiting the southern portions of the Colorado Plateau and the Four Corners region from Late Basketmaker times until about 700 years ago. *Anasazi* is derived from two Navajo words meaning enemy ancestor, ancient enemy, or words to that effect. Something is lost in translation, but it does illustrate the potentially hostile relationship that can exist between the mobile, semi-nomadic hunter and the sedentary villager growing corn and squash. Today the Hopi, Zuni, and other modern Pueblo People find the old term *Anasazi* offensive and it is no longer used by scholars, professional archaeologists, or the National Park Service. Today the preferred terminology is Prehistoric Puebloan People.

**COHONINA CULTURE**

The Cohonina were a non-Puebloan People, who inhabited the Coconino Plateau from the San Francisco Peaks to the South Rim during Puebloan times. Their signature pottery is known as San Francisco Mountain Grey Ware and it is one of the several material traits distinguishing them from purely Puebloan culture. Today the diffuse remains of a Cohonina settlement in the area of Pasture Wash west of Grand Canyon Village along the South Rim intermingles with sites that are purely Kayenta Puebloan. Artifacts produced by both cultures are found on both types of sites—Cohonina arrow points on Kayenta sites and Kayenta ceramics on Cohonina sites. This is a strong indication there was a period during the eleventh and twelfth Centuries when relations between the two overlapping groups were amicable and mutually advantageous. When the drought came, after 1180 A.D., their sites were already deserted. There is some ethnographic evidence they were Yuman language speakers and may have ultimately been absorbed by the Hualapai Confederation moving up river and they morphed into the modern People inhabiting Cataract Canyon (Supai).

summer rains and the opportunity of new lands to the west just made the move all the sweeter.

Successful traders and craftsmen, the Puebloans, like all tillers of the soil, were calculated opportunists. Farmers are conservative by nature, and risk takers by necessity, balancing what they can grow with the quality of soil and the availability of water; a slight miscalculation, and the very young and old could die. Two bad years in a row and even the strongest are staring at an empty plate. Farmers are always thinking ahead and taking advantage of subtle changes in the environment. This is what happened at Grand Canyon by 1040 A.D. The farmers recognized an opportunity and expanded into the canyon like water pouring into a dry stream channel. Carrying their infants, bows, water jugs, a few turkeys, and seed, with small children and dogs in tow, they moved westward from their old homes in south eastern Utah and the Four Corners area. We know there were already a few Puebloan (Uto-Aztecan speaking) families along the river, possibly blood relatives, and Cohonina (possibly Yuman speakers) living on the South Rim, but it was still a frontier and for the time being there was room, water, and soil to spare. Within a generation the new colonists had occupied virtually every arable delta and useful side drainage in eastern Grand Canyon. Think of Indiana, Illinois, and Iowa from 1820 to 1850, or Iceland from 870 to 970 A.D.

Corn was the bulwark of Puebloan life, being central not only to the table, but to the daily fabric of the community. However, a steady diet of corn can lead to a variety of ailments stemming from a lack of B-complex vitamins and certain amino acids. The high starch content forms sugars that wreak havoc in the mouth, causing tooth decay and painful life-threatening abscesses not seen in groups where wild foods dominate the diet. This serious

## AGAVE, MESQUITE, AND ROASTERS

As the canyon cuts westward and the river drops in elevation, the resources available to the marginal farmer and the hunter-gatherer alike change. Below Middle Granite Gorge, agave and mesquite become important as a food source.

In the canyon's west end, 1,000 to 700 years ago, the Virgin Puebloans were roasting agave as a supplement to their corn-based diet. Possibly the most common archaeological sites in Grand Canyon are agave roasting pits. The river-corridor site sketched here was used off and on for more than nine centuries, right up until the 1900s, by successive generations of Virgin Puebloans, Cerbat, and Southern Paiute families.

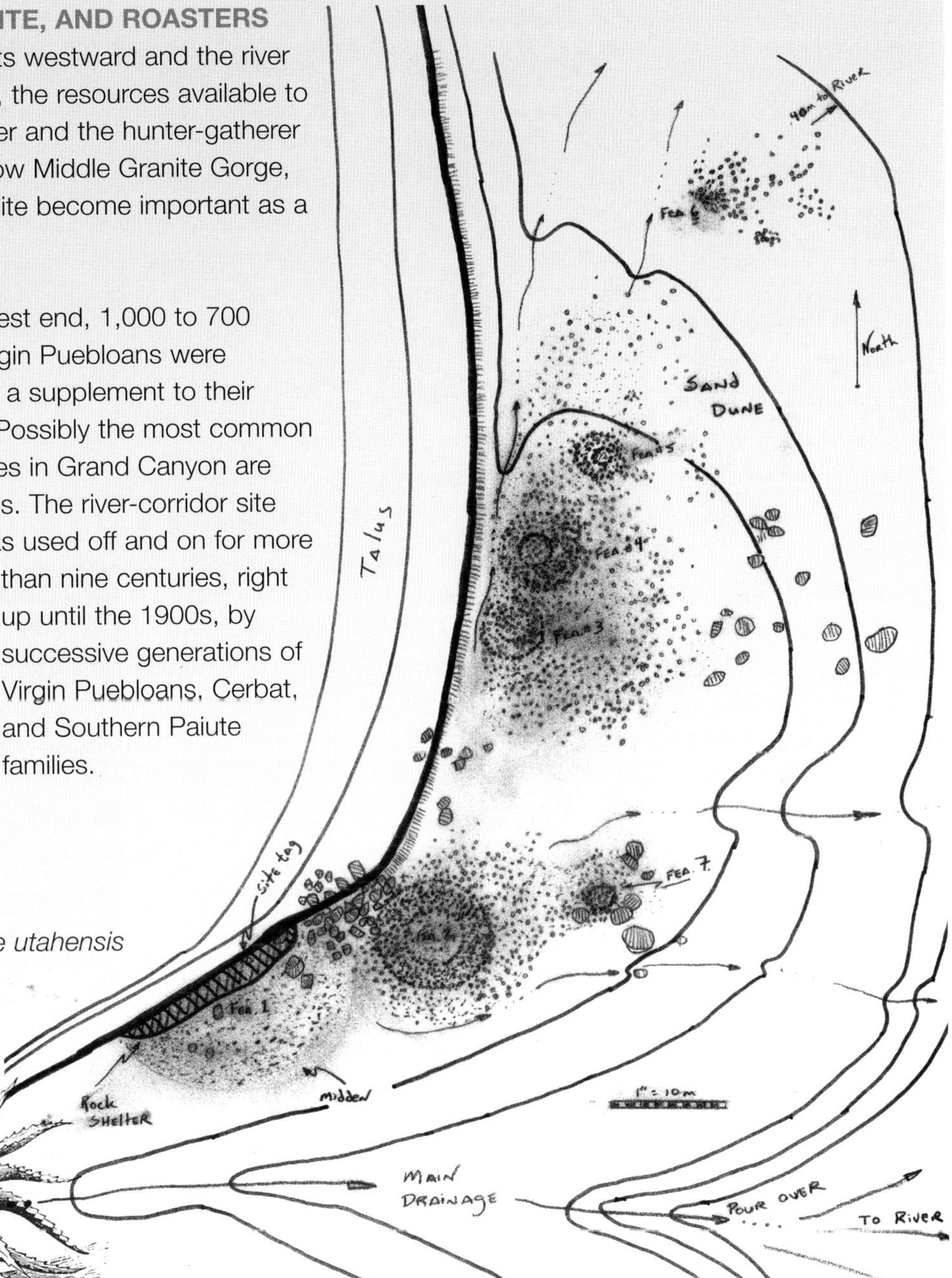


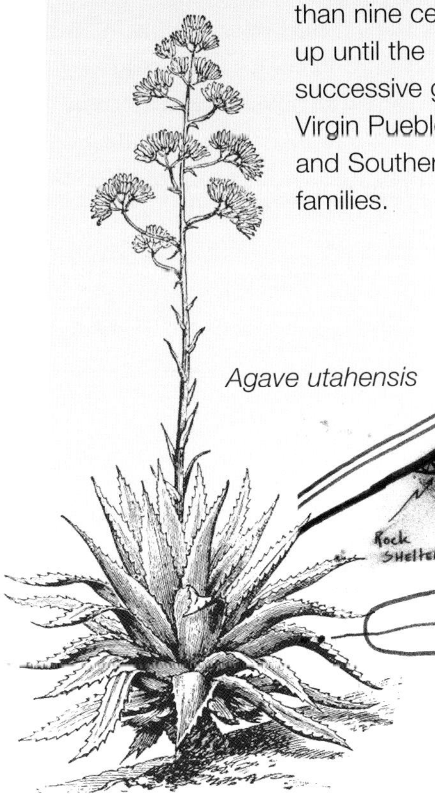

*Agave utahensis*

problem can be alleviated by soaking corn in ashes (which contain lye) to turn it into vitamin-B-rich hominy. If corn is complemented with squash, beans, and locally collected greens, diseases of nutritional deficiency can be averted. This New World triad of corn, beans, and squash provides the carbohydrates, protein, amino acids, and vitamin A that the body requires for growth, brain function, and healthy eyesight, all particularly important to early childhood development.

Despite the new lands and plenty of water, the people could not afford to be just farmers. The severity of the Grand Canyon system would not allow it. Even with rich alluvial terraces, increased precipitation, and a higher water table, which are all gone today, farming was still risky business. So to hedge against a bad harvest or a severe flood in addition to farming, they capitalized on the natural resources available to them. When the menu can be supplemented by wild plants and animals, everybody is much better off. Prickly pear fruit, lemonade berry (*Rhus trilobata*), organ meats, and rose hips are rich in vitamin C, and pinyon nuts are high in good fat and protein. In fact these families probably relied substantially more on pinyon nuts they gathered than beans they grew. Yet even pinyon pine production waxes and wanes from year to year and place to place. Because of their broader resource base, Puebloans living at Grand Canyon may have had a nutritional advantage over their urban cousins at Keet Seel in northeastern Arizona or Poncho House in southeastern Utah.

In Grand Canyon, useful things were stacked one on top of the other for a vertical mile, from the river to the rim. There were in this vast arid country edible cactus, mesquite

**THE MOMENTOUS ELEVENTH CENTURY**

Across the world and within the region, this century was a time of change. The Hohokam, Fremont, and Mogollon cultures were flourishing. The multiple eruptions of Sunset Crater, outside of Flagstaff, in the 1060s instigated the earliest wave of Sinagua settlements in the area where Wupatki would be built eighty years later. On the southern Great Plains, Native Americans were moving into villages and growing corn. In Scotland, Macbeth was slain by Duncan in 1050. In 1066 the Normans invaded England and were on the offensive throughout Europe. It was the Fugiwara Epoch in Japan and the beginning of the cult of the Samurai. The Polynesians were expanding eastwards across the Pacific colonizing Hawaii and Easter Island. The Toltecs ruled central Mexico. Cahokia, along the banks of the Mississippi River in what is today Illinois, was the largest city in North America.

pods, yucca, agave (mescal), grass seeds, acorns, walnut and pinyon nuts, wild fruit, greens and herbs, and plants used as medicines, dyes, and for ceremony. Yucca had multiple uses: the fruit for food, the tough fibers for cordage and sandals, and the root for soap as well as pliable red roots woven into basket designs. Animals utilized included bighorn sheep, deer, bear, bobcat, mountain lion, rock squirrel, mice, packrats, woodrats, eagles and hawks, waterfowl, chuckwalla, and small lizards. Like many southwestern tribes, the farmers were apparently, by choice, not fishermen.

There were raw materials in abundance: wood for fuel, chert for stone tools from Kaibab and Redwall Limestone, cobbles for hammer stones, multiple grades and textures of sandstone for manos and metates, valuable mineral pigments, turquoise (malachite and azurite), green calcite and serpentine for jewelry, good clean salt precipitating out of the 575 million-year-old Tapeats Sandstone, soil, and most important of all, water.

## LIFE ON THE DELTAS

In the eastern Grand Canyon are a series of large side canyons that drain into the Colorado River. These tributaries breach the incredibly rugged terrain existing between the forested rims and the seemingly desolate inner canyon. Acting as the routes of daily life, the side canyons were the highways by which the inhabitants accessed the stair-step ecology of Grand Canyon.

Each of these side-canyon systems creates a large delta at river level suitable for farming and village life. The deltas served to focus settlement and because of that they are often the focus of the

Unkar Delta on the Colorado River is visible from Walhalla Plateau. The difference in elevation from point to point is about 5,200 feet.

archaeologists. The big canyons—Nankoweap, Kwagunt, and Unkar—drain into the Colorado from the north, and the Palisades, Tanner, Cardenas systems from the south. Several secondary side canyons such as South, Basalt, Sixty-Mile, Chuar, and Fossil Canyons had less arable ground yet were still workable and were also occupied by Pueblo families. In those days an extensive system of alluvial terraces also existed in the river corridor, adding considerable ground that could be cultivated. These terraces were the connective tissue along the margins of the river; most of them are gone today along with the sites they hosted. (Anybody who has driven through Tsegi on the Navajo Reservation and looked down into Tsegi Wash can see where Navajo families have gardens and small orchards precariously perched on the erosional remnants of a once larger terrace system).

Large ollas, such as this corregated vessel fitted with a fiber lattice for carrying, could hold as much as twenty gallons of water.

The delta farmers of Grand Canyon were growing as much as they could for as long as they could, possibly even getting in two overlapping harvests in good years closer down along the river. They double-cropped to whatever extent possible, farmed both the inner canyon(s) and the rims, and took advantage of calories found in native plants and game within their range when necessary. They stored food for winter and just as critically for any eventual drought. Below the rims in the lower elevations of the canyon's western reaches, early spring would bring agave (mescal), and edible greens would be popping up along the river. By April people could gather a variety of edible plants where the canyon opens up. As soon as the time was right, squash, cotton and probably some beans were planted. On the rims, corn sown in late spring matured through the early fall and the upland harvest would dovetail nicely with the ripening pinyon nuts and the best time for deer hunting.

The strategy adopted by the delta farmers in the pursuit of an adequate diet is best illustrated by the geographic relationship between Unkar Delta, elevation 2,600 feet above sea level, and Walhalla Glades on the North Rim, at 7,800 feet. Unkar Canyon, which connects the two, allows access to the often snow-covered pine forest of the Walhalla Plateau from the delta at Unkar in just a few hours on foot. These colonists were well connected by social obligation, blood ties, and physical trails. Even though their farmsteads and small villages may have been geographically isolated, they did not live in isolation. Like all people living on a frontier, they relied on each other. There was no one else to rely on.

The Walhalla Plateau can get more than 200 inches of snow per year. Because of the high elevation, snow can still be present on the ground through May. The moisture derived as the late snowpack melted was held by the soil and allowed the seeds planted in late spring to germinate. The porous and highly fractured Kaibab Limestone that caps the canyon rims as a rule does not permit water to stand on the surface, so there were no ponds, streams, or springs to rely on. Any moisture from the late snow that the fragile soils can retain is critical to the survival of a rim crop until the monsoonal rains come in July or August. The large, locally made ceramic vessels known as ollas, some of which will hold about 20 gallons, could be packed with snow and kept in the shade for watering the young plants in the critical time

Remnants of seasonal dwellings and walled gardens are found on the Walhalla Plateau. It is unlikely anyone wintered over during Puebloan times.

## GRAND CANYON THERMALS

Anyone who has spent time in the river corridor or camped on the edge of the canyon has experienced the exchange of warm air with cold and vice versa. It may be good for high-altitude farming, but it makes wintertime river trips just that much more uncomfortable. Here is what happens: Most people start getting up around six a.m. It is pitch black. Most of your gear is damp. Within the hour the pale light comes. You cannot see the sun, but it is drawing heat upward, and cold air is being sucked down from the high country, channeled through the side canyons to the river, spewing a tide of frozen air out across the deltas. Within minutes, the temperature drops eight, ten, sometimes fourteen degrees freezing the damp gear.

The survival of Puebloan farmers depended on understanding this phenomenon. Knowing where crops were harmed and where they would benefit from the convection was critical. For the modern river runner or archaeologist it might mean the inconvenience of cold feet, but for the canyon's inhabitants 900 years ago it was the difference between a bumper crop and starvation.

## PREHISTORIC PUEBLO CERAMICS

Ceramics are crucial in confirming dates for habitation sites. Archaeologists classify ceramics based on type of clay, firing conditions, temper, color, technique of manufacture (coil and scrape, paddle and anvil), paint (mineral or carbon), and surface design. Once classified as a recognized "ware," vessels are further assigned to types and styles. For example, Tusayan White Ware, a Kayenta Puebloan type found in Grand Canyon, evolved over time into various styles of painted design.

Often wares have unique attributes. Crushed olivine temper from volcanic rock makes broken sherds of Virgin Moapa Ware seem to be filled with tiny emeralds. Manganese-rich clays from the canyons east of Kanab, Utah, give Shinarump Wares a distinctive purple cast.

Ceramic types have variable duration in the archaeological record. For instance, Black Mesa Black-on-White was produced without notable change for 300 years. On the other hand, a type of Tsegi Orange Ware known as Medicine Black-on-Red was made for only about 50 years.

We largely define the prehistoric Pueblo by their pottery. Beautiful black-on-red, black-on-white, polychrome, and plainware vessels were produced, traded, and used on a large scale for several centuries. Formal decorated wares were highly valued and extensively traded, while corrugated plainware was made in each community for everday use in storage and food preparation. Sometimes whole vessels were buried with individuals.

People repaired cracked vessels by drilling small holes on each side of the crack and stitching it with yucca or hair cordage. They recycled sherds of broken pots by grinding them smooth and shaping them into pendants, spoons, scraping tools for wet pottery, and spindle whorls for cotton yarn. They used sherds to chink walls of the pueblo and ground them up as temper for new pots.

Tusayan White Ware, like clothing fashion or car designs, evolved over time, into various styles of black painted design. Identifying these can help frame a specific site into a window of time.

KANA-A – A.D. 725-950

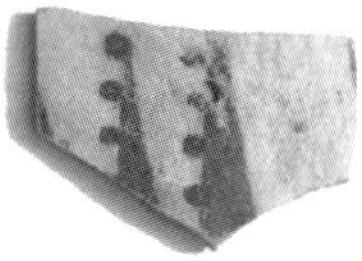

BLACK MESA – A.D. 875-1150

SOSI – A.D. 1050-1200

DOGOSZHI – A.D. 1075-1200

FLAGSTAFF – A.D. 1100-1275

prior to the coming of the monsoon rain. The farmers also scraped shallow cisterns into the limestone and lined them with clay in order to pool crucial water. But the presence of adequate soil and a sustainable supply of water were still not enough to guarantee a crop.

Temperature is the third factor critical to growing corn and it seems unlikely that Walhalla was warm enough for long enough. What really allowed for this high and windy spot to yield a crop was the unique circumstance of the spot itself. Enter the grand design of Mother Nature. The Puebloan sites at Walhalla Glades are perched on the edge of an abyss. During late spring and early fall, when most locations in the surrounding countryside can experience a sudden killer frost, the glades are bathed through the night and on into early morning by great drafts of warm air climbing the vertical walls of the canyon, surging over the rim onto the gardens. Corn and vine crops do best when allowed to mature in conditions where nighttime temperatures exceed 50 degrees Fahrenheit. This warm air convection allowed the natural sugars within the plants to set, causing larger and tastier yields.

Among its many ecological complexities, the Grand Canyon is also a massive breathing device. Just as the canyon walls direct the warm air upward, the intricate web of drainages and side canyons funnels cold air down into the main canyon through the late night and early morning hours. Nightly temperatures can vary as much as 12 degrees between a slope and its companion drainage only a hundred yards away. For fear of killing frosts, most habitation and agricultural sites on the North Rim are not found in valley bottoms where the chilled air settles and where soils might be richer, but are on the gentler slopes closer to the rapture-inducing cliffs.

The gardens at Walhalla Glades would have been busy places from late spring through fall hunting season, but the more than 200 inches of snow per year would have made winter habitation miserable, if not impossible. Nobody

| LANGUAGE FAMILY | LANGUAGE GROUP | LANGUAGE/PEOPLE |
|---|---|---|
| Nadene | Athapaskan | Navajo/Apache |
| Uto-Aztecan | Numic | Southern Paiute |
| Uto-Aztecan | Hopic | Hopi |
| Hokan | Yuman | Hualapai/Havasupai |
| Zuni | Zuni | Zuni |
| Kiowa-Tanoan | Tiwa | Taos |
| Indo-European | Romance | French/Italian |
| Indo-European | Germanic | English/Dutch |

Languages change at predictable rates over time, allowing linguists to determine when certain groups of people diverged from each other. Like countries and buildings, languages accommodate widely varying numbers of people. Nadene, for instance, includes hundreds of thousands of Navajo and other Apache speakers. Only about a thousand people fluently speak Zuni, which is its own exclusive language family.

wintered over that high up, especially in view of the fact that the sunny desert was a few hours walk downslope and travel to the south and the lower elevations downstream to the west was easy. This makes even more sense when taking into consideration that the river in winter would have been low and more easily crossed. Snow is uncommon at river level, but the temperature can and does regularly fall below freezing. During the archaeological survey of 1990–1991 our rubber rafts would occasionally freeze to the clay on the beach overnight.

Most of the delta farmers would have been found along the river corridor and in the bench country below the rim after the first snows through spring planting. After spring planting on the deltas and terraces of the inner canyon, families farming the rims would have moved upward as the snows melted, prepping their plots, upgrading terraces, and repairing check dams, making sure the clay-lined reservoirs and ollas were full, getting everything ready.

Conversely, the inner canyon is very hot from May through September, so any reason to be on the rims during summer would have been sufficient. Despite the heat some family members were needed along the river during summer to tend the gardens, water the maturing plants, keep the rodents and ravens at bay, and pray for rain.

Farming was not only a physical effort, but a spiritual and religious exercise as well, requiring cooperation. Life was not partitioned into work, home, church, and lessons. It was one package. Individuals as well as families considered themselves extensions of the local community, not separate from it. The clan relationships which perpetuate order and enforce the unwritten social regulations in small close-knit communities are extensive and include defining acceptable behavior, food taboos, who your cousins are, hunting prohibitions, selecting an appropriate marriage partner—which because of your clan affiliation might require a trip of 20, 30, or 50-plus miles to make official—and more critically along those same lines, whose side you're on in a serious dispute. Farmers may or may not be more devout than one group or another, but they do tend to be meticulous about tradition and usually incorporate devotion into the ritual of daily life, including caring for the growing corn.

Farming communities do what is required in accordance with the rhythm of tradition and in response to the dramatic events beyond human

### LIFE EXPECTANCY

Accepted data for prehistoric Puebloan longevity from old studies indicates average life spans of twenty-nine to forty years. This age is artificially low, brought down by infant mortality and high death rates for children less than seven years old. In fact, a singular lack of infectious diseases in the New World prior to European contact would, barring starvation and trauma, tend to make for an overall healthy native population. However, studies tell of a hard life for Pueblo people: lower leg breaks associated with desert farming, episodes of severe malnutrition, debilitating arthritis, teeth often worn down to the gum line, abscesses, and the maladies of hard work.

The well-known granaries at Nankoweap. Storage granaries are a common site-type in Grand Canyon. Air tight, water tight, and out of the sun, they preserved caches of corn and other dried foods through the winter and were insurance against crop failure, insects, mold, extended drought, and hungry rodents.

control: unusually big floods, rainless monsoons, raiding parties with better weapons, bumper crops of insects. Tactics used by the farming settlements throughout Grand Canyon were fine-tuned to their particular home bases. The Kayenta farmers in eastern Grand Canyon, the isolated Virgin colonists at Pinenut near the North Rim west of Kanab Creek, the several families along Tapeats Creek to Thunder River and above Deer Creek Falls, and the numerous Kayenta delta farmers would have had schedules as well as rituals specific to their own needs.

As the weather allowed, individuals and groups spread out within their resource zones as necessitated by the demands of the season. Life required simultaneous attention to several tasks in several places throughout the year: tending crops, watching small children, processing corn, hunting, making new pots and fixing old ones, gathering materials for baskets, collecting wild foods and plant medicines, repairing granaries, feeding the turkeys, visiting, haggling with traders, keeping the faith, and keeping watch. Everyone had a job to do. Very little was wasted.

And people died. Especially in a small group dynamic like a farming village or an arrangement of allied families moving across the landscape, the loss of anyone through death or incapacitation in any number of ways is felt by everyone. Their crucial knowledge and practical skills could be lost with them, unless they were already passing it down to a younger family member. Or in other critical areas such as spiritual and ritual knowledge or important songs, if not transferred to someone they deemed worthy, provided somebody worthy was there, those songs could be lost too. Whatever the cause, an untimely loss creates a hole to be filled and that loss can be multiplied through famine or flood; recovery was no sure thing. That is why broad-based traditional knowledge at the family level and continuity over time is the cultural bottom line—that is why a lot of what used to be still remains alive today at the family and clan level within modern tribes.

Tusayan Pueblo as it may have appeared around A.D. 1185.

## ARCHITECTURE

The Puebloans built a wide array of permanent stone structures, roofed with wood beams, cross-hatched by branches and sealed with clay, as well as expedient seasonal ramadas (formal shades) and shelters of post and brush. A variety of configurations exist throughout Grand Canyon: linear blocks of rectangular rooms, D-shaped and bow-shaped pueblos with attached plazas, L-shaped configurations with wind breaks, detached circular rooms, and stone partitions placed in overhangs. For more expedient and temporary structures they also built features out of wood and jacal (like adobe), but the clay melts and the wood rots so these have been poorly preserved in the record compared to stone.

This variation reflected personal preference, local conditions, available materials, even clan affiliation, and illustrates at a very personal level that Puebloan civilization was anything but homogeneous—remember that the vibrant colors we would have seen in those days are long gone. Yet patterns and faded colors remain. Main entrances were open to the east regardless of the orientation of the structure to the river. Between 1040 and 1190 A.D. the average size of a room grew from a tight 9 square meters to a luxurious 15 square meters. More than any other statistic, this increase shows how successful these folks were in a tough rocky place. They were growing and gathering more food and having more babies so they needed more structural space for larger families and greater storage capacity.

These people spent most of their lives outside, going indoors because of bad weather or for a specific activity, like telling winter stories or conducting a healing ceremony. Their idea of personal space would differ from a modern Western one. No one had their own bedroom. There was one bathroom: outside. At any rate, finding personal space was not likely a problem at Grand Canyon. Anyone who required a little solitude (Holy Men in particular) was quite aware that they were living in Solitudes Cathedral and only needed a short walk to be in a private place without the necessity of retiring like a medieval monk to a small, poorly lit room.

A farmer's worries are never put to rest. The harvest must be secured and protected, not only from those who might come and take it, but more poignantly from the enemies living among the people themselves; rodents, insects, moisture, and sunlight. Corn, squash, prickly pear fruit, meat, agave/mescal, a variety of seeds, and pinyon nuts would be processed, dried, and stored for winter.

Once the families' needs were covered sufficient seed was counted out for the next planting and any surplus was prudently set aside for trade, social events, and that bad year (or two) always looming on the horizon. Within a pueblo community much of the structural space was devoted to storage. At the isolated Pinenut site situated near Water Canyon back from the North Rim, a small living room accommodating the entire family was attached to a linear string of smaller storage rooms. Here, 70 percent of the square footage was devoted to storage space. More than 75 percent of the

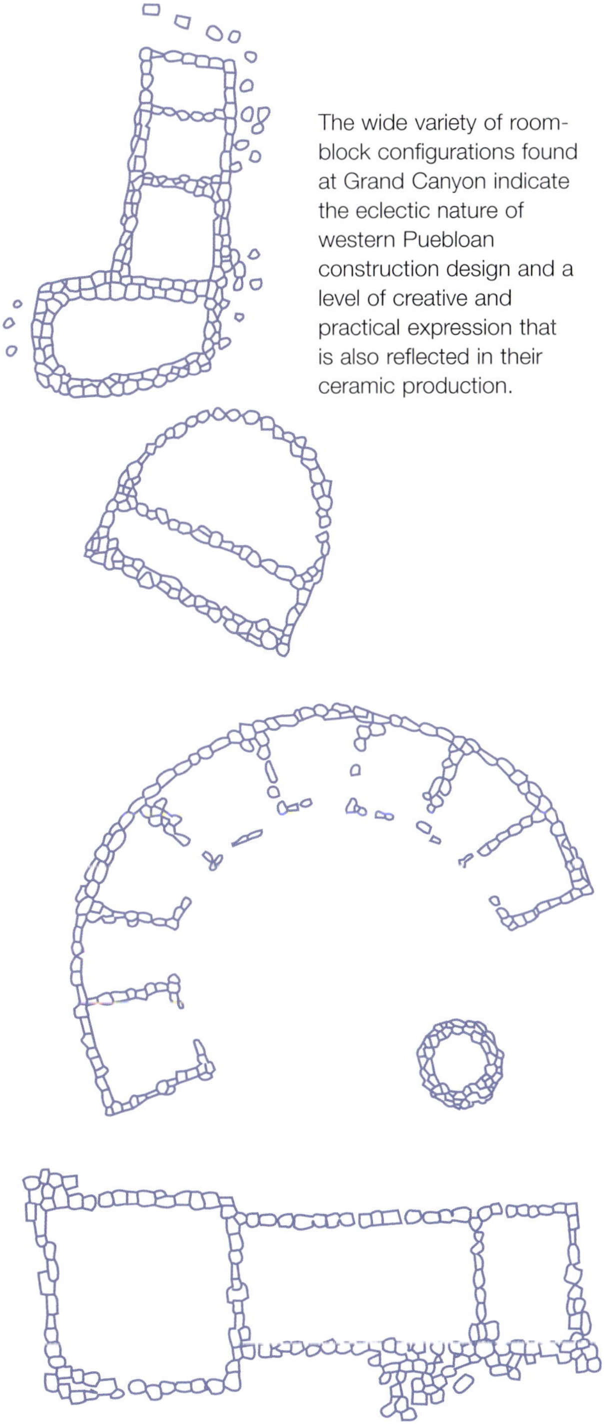

The wide variety of room-block configurations found at Grand Canyon indicate the eclectic nature of western Puebloan construction design and a level of creative and practical expression that is also reflected in their ceramic production.

**PITHOUSE TO KIVA**

The pithouse was the preferred domicile of Southwestern peoples from Late Archaic through Late Basketmaker times. As the early Puebloans abandoned the pithouse as a living space, some, but not all, retained it as the kiva. Some kivas were of square design to accommodate local space and materials, but all incorporated the basic elements — a sipapu, firepit, ventilator shaft, and roof entry. Large kivas were being constructed along the San Juan River in southeastern Utah about 1,300 years ago. The kiva is central to many modern-day pueblo societies.

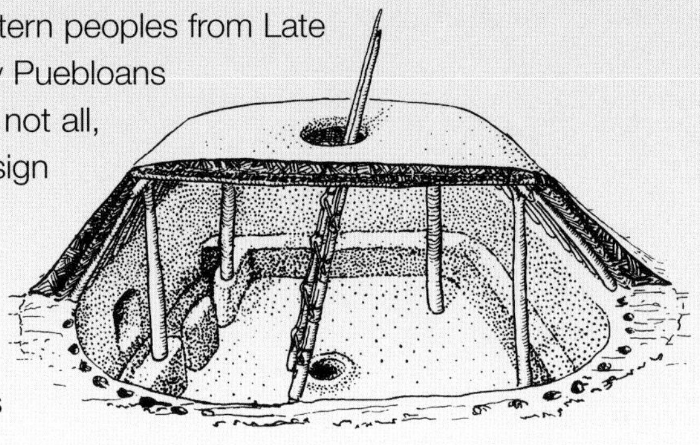

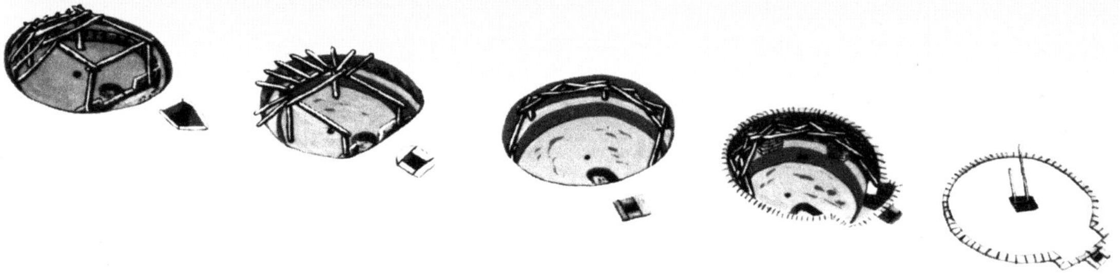

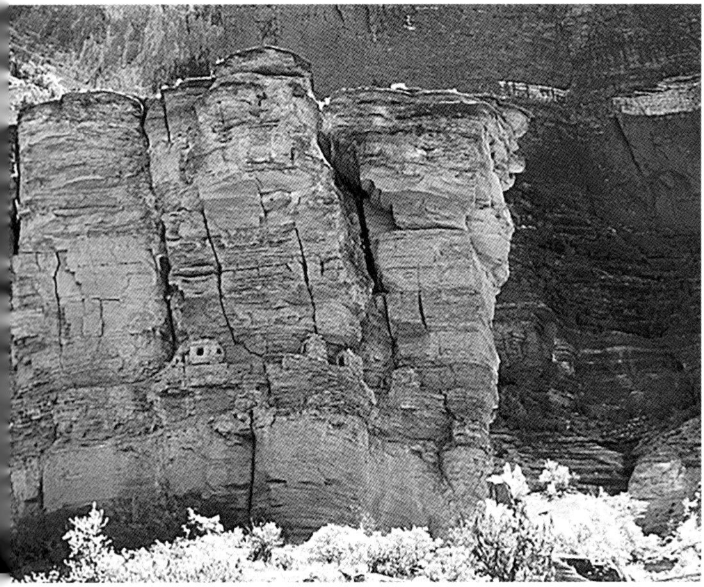

Cliff-face granaries in Clear Creek Canyon.

2,000 pottery sherds recovered from this site were from broken storage jars.

After the home capacity was used up, people built clusters and individual granaries in the cliffs. They found a suitable location, walled it off with stone and clay, and then meticulously fitted the opening with a wooden or sandstone slab door. The impressions of corn cobs and little hands—presumably children's—are often found on the finished surfaces. This makes sense as the doors are generally small, only big enough for a child to get in and out. This tells us that, as in any farm family from any century, you became an active part of the team at a young age and were personally invested early on in a life-long regimen.

As mentioned earlier, prior to the use of above-ground pueblos Basketmaker people lived in subterranean pithouses. By the ninth century farmers were opting more and more for the surface roomblock, and the family pithouse was often converted into a kiva. The Fremont culture to the north in Utah never abandoned the pithouse until they abandoned the entire region. This is also true for some Sinagua families in central Arizona. The first known large ceremonial kiva is in Bluff, Utah on the San Juan River, a tributary of the Colorado. A lot of culture moved from San Juan country downstream into Grand Canyon—people, redware pottery, and probably the old Puebloan religion.

Once hearth and home to the entire family, these circular underground rooms flipped and became the sole domain of men and the epicenter of spiritual, ceremonial, and economic forces within the Kayenta, Mesa Verde, and Chacoan traditions. Northeastward into the old Puebloan heartland, kivas were commonplace. They are less common at Grand Canyon. Kiva religion is virtually absent from the Virgin and Cohonina traditions, nor was it practiced by the Fremont or Sinagua. The Hopi and Zuni along with many other modern Pueblo societies still practice the old religion.

Back at the Pinenut site on the North Rim west of Kanab Canyon, when the new stone-walled pueblo was constructed, the old pithouse was not transformed into a kiva. After it was

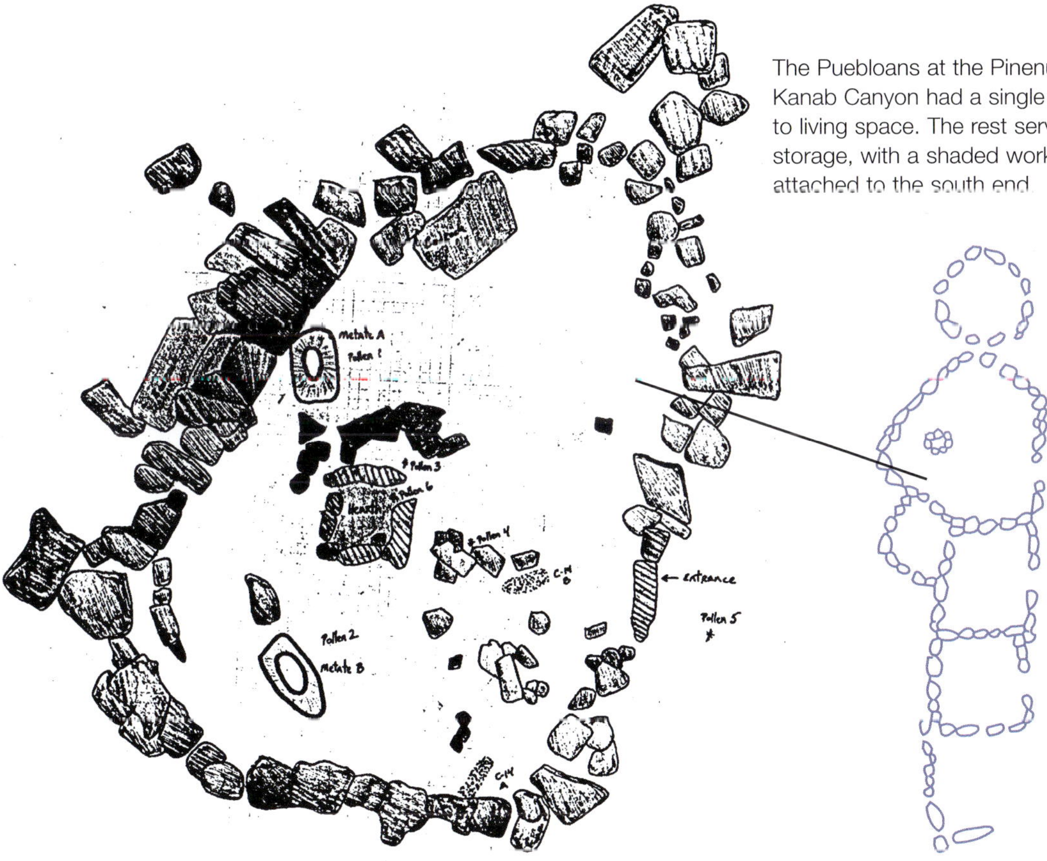


The Puebloans at the Pinenut site near Kanab Canyon had a single room devoted to living space. The rest served as storage, with a shaded work area attached to the south end.

abandoned the resulting open hole was used by the family for a while as a trash "pit" and used later still as a garden plot. The piety of the old Kiva religion, which spread outward from what would later become southeastern Utah, may have been considered just so much inconvenience to the Virgin farmers working the dry uplands farther west. The Cohonina would have had their own religion. It seems according to the evidence that they were not growing much if any corn at all, so they may not have been focused overly much on praying for rain and had no use for a formal ceremonial structure like a kiva.

It is also possible that some of these hard-scrabble frontier farmers used above-ground structures that are no longer identifiable as having been used for religious purposes; this could include perishable sweats and/or open air communal dance grounds in special places removed from their daily lives. The enigmatic shoe box–shaped stone walls overlooking the river at Cardenas Canyon, mistakenly dubbed by

## TUSAYAN RUIN

By the late 1100s prehistoric Puebloans in the canyon region were "grouping up" into extended communities rather than living on family farmsteads. The pueblo we call Tusayan, sitting just one-quarter mile south of the canyon rim near Desert View, is a good example of community land use in an arid climate. Park archaeologist Jan Balsom says "... the location was excellent. There were neighbors, nice views of the sacred peaks, good farming in the drainage below the site, and a wide variety of options for hunting and gathering. A trail (known today as the Tanner) into the canyon, where subsistence activities could take place, was a short walk away. For trade, the Coconino Basin and rim areas could be easily approached from the pueblo."

Tusayan was not built until about A.D. 1185, almost simultaneously people had begun to withdraw from the inner canyon in earnest. After twenty years of occupation, the pueblo's thirty or so inhabitants joined the migration south and eastward, leaving their homesite as a testimonial of their movement to a place of permanence.

Hilltop Fort at Cardenas Canyon.

Below: Serpentine pipe fragment from the Cardenas area.

white miners as the Hilltop Fort, may have been such a location. Or it may have been a way-place for Hopis trekking to and from the Grand Canyon to trade with Havasupais or collect salt in the inner gorge.

A formal kiva can still be seen at Bright Angel Pueblo, the same site observed by Major John Wesley Powell in 1869. It is situated just downstream from the Black Bridge that connects the south side of the canyon to the facilities at Phantom Ranch. This pueblo is geographically isolated deep in the granite and perched directly along the river, making regular daily treks to and from other villages inconvenient. This site was oriented tactically by terrain following Bright Angel Creek up-canyon to Ribbon Falls and the North Rim and over the rock upstream along the Colorado River to the sites at Clear Creek. The farming could not have been good on this small rocky delta, with limited sunlight and prone to flooding from Bright Angel Creek as well as the main trunk of the Colorado River. It was occupied off and on for over a century, but for some reason it was an important enough place to have a formal kiva and men were coming here routinely for that reason alone.

## GARDENS

Agricultural features common to southwestern farming—terraces, ditches, shallow clay-lined depressions, check dams, and actual garden plots outlined with rock borders—are found throughout the canyon country: Walhalla Glades, Kwagunt Delta, Deer Creek, Pinenut, Cataract (Havasu) Canyon, Tusayan, Pasture Wash, and Basalt Canyon, to name just a few. These features were constructed and maintained to channel and store water as well as to hold the fragile soil in place.

Protected and nurtured, every plot was treated with due reverence as the means by which the farmers could stay in the canyon. Each

Prehistoric Puebloan farmers used low rock walls to slow erosion. Today the National Park Service uses a Zuni check dam team to construct wicker work and rock dams to slow erosion at archaeological sites in the river corridor.

plot was surrounded by a low rock wall that acted as a passive solar device, preventing frost damage to young plants along the river in early spring and in the uplands in early summer. The walls also allowed better retention of moisture and kept organic material from dispersing. Modern Pueblo farmers often plant faster-growing wild food plants with domesticated crops to act as shade for the young seedlings. Amaranth is used even today in this way by the Hopi.

If the canyon can be defined in any singular fashion, it is as a monumental engine of erosion. Slowing even the smallest increment of erosion is an act of human defiance in the face of the inevitable. Ultimately, the canyon wins this battle. Nevertheless, maintenance of soil is key to successful farming at Grand Canyon—no soil, no corn. So the old Puebloans spent a lot of time tending to their plots. When the soil began to erode quickly in significant amounts after 1170 A.D., quite notably so did Puebloan culture.

## WATER SOURCES

Every water source — seasonal seeps, springs, bedrock tanks — would have been part of each person's mental map, along with trails and every other important resource. Water knowledge made up part of the unwritten cultural lexicon known and understood by each member of the community, young and old alike. Dying of thirst in Grand Canyon is a modern phenomenon attributable to the inexperienced, the unskilled, and the unlucky. It is very hard to imagine a Puebloan, Hualapai, or Paiute dying of thirst.

## TRAVEL AND TRADE

Commerce was an important facet of life known to all the peoples at Grand Canyon, linking them economically and of equal importance, socially, with their blood relatives and neighbors and the world beyond. As the weather in the uplands warmed and the water level in the river dropped after the spring runoff, cross-canyon movement—so essential for daily chores and social connections and trade within and outside the canyon—would resume. Travel through the inner gorge along the river is difficult at best. Efficient travel up and down the river takes place high above the river on the Tonto Platform and Esplanade, which can be accessed by side canyons. An intricate system of local trails existed in the canyon long before the backpacking craze of the twentieth century. Modern trails mimic old prospecting routes that followed still older native routes.

Many old trails can still be seen. The trail originating on the bench above Basalt Delta linked this settlement with larger villages downstream at Unkar Delta and Furnace Flats. Others are the old Hopi Salt Trail, the trail to a hematite mine below Parashant Canyon, and the maze of trails traveled by the Hualapai and Southern Paiute. When Hopi guides led Spaniards to the edge of the abyss in 1540, it was on purpose to a place where no trails were visible from the rim, the Salt Trail in particular.

Coastal shell beads were one of a variety of trade items brought to the canyon.

The Anasazi Bridge

People expedited local travel by taking direct routes from point to point. On the north side of the river, upstream from President Harding Rapid, are the remnants of a Puebloan footbridge constructed from driftwood sticks and spanning a gap high in the cliff face. This precarious structure allowed local farmers to scale a thousand-foot bluff and get to the rim quickly. (Today no access is allowed to the site. Known as the Anasazi Bridge, this remarkable feature can be seen from the river.) Other such examples were observed by U.S. Army expeditions of the 1850s in the west end of the canyon. Simple forked-stick ladders connected ledges and systems of benches that would otherwise require several hours to walk around. Archaeological field crews on the 1990–1991 survey

Basketmaker II feather fan from Marsh Pass, Arizona

encountered a lot of these stick ladder features below Lava Falls, where trackless rock ledges and basalt flows dominate the inner canyon. A simple stick ladder in the right spot could save a several mile walk around.

At first glance the river seems to be an insurmountable barrier, but in actuality it is only a formidable obstacle that people overcame. The exact method of the wet commute is not known, but it apparently happened without much trouble when the conditions were right. Animals do it daily. Deer are commonly seen making the crossing. On one of our work trips a three-foot rattlesnake was observed in mid-channel two miles above Lava Falls, bee-lining for the north bank. On another trip we saw a lizard cruising downstream on a chunk of driftwood. During many years, water levels would have been low enough to cross on foot from July or August through the following March. Except of course during other years, when late snowmelt in the Rockies created prolonged runoff, or intense monsoon rain caused unpredictable summer flooding, options for crossing the river would have been more limited. Prior to the operations of Glen Canyon Dam the water in the river would have been much warmer during the summer and fall than it is today and that would have been a big factor for a person crossing on a log.

Trade items coming into the canyon were exotic and diverse: complete ceramic vessels from the east, shell bracelets and slate beads from the south, shell beads from the Pacific Coast, and projectile points made by the Fremont. Routes of trade ran upstream to southern Utah and the drainages of the San Juan River. Redware pots made over a thousand years ago in the vicinity of Bluff, Utah, are found in Grand Canyon. Shell bracelets made by the Hohokam in far southern Arizona were shaped from mollusks that had lived in the Gulf of California. Macaws from Mexico are found as far north as Westwater Canyon in southeast Utah, carried by traders walking Puebloan roads.

Before the reintroduction of the horse to the American Southwest by the Spanish in 1540, all American Indians—not just the people of the Grand Canyon—traveled on foot. Oral histories

Field rendering of pigments in pictograph.

On an obscure ledge overlooking Lava Falls Rapid there is an unusual pictograph panel consisting of three Kachina figures. It is curious because the Pueblo people had for the most part moved out of the canyon before the Kachina religion made its way onto the plateau. This panel reminds us that small groups of Hopi traders, were returning to this particular place in the canyon. They had moved on, but old ties and new business were bringing them back.

tell of migration centuries ago when the Arikara coming down from the north encountered the vast Grand Canyon before heading east over the Rockies and out onto the Great Plains. The Hopi culture hero Tiyo floated on a log down the Colorado River through the canyon to the sea, returning to the Mesas with new songs and instructions for the Snake Dance. This is another example of an oral tradition reinforcing the connections between the Uto-Aztecan speaking Puebloans of the eastern Canyon to their Yuman speaking neighbors downstream. There are many such tribal stories. Being on foot did not hinder the rapid movement of people and goods over vast distances. The Hopi would trek for three days from the Mesas to Supai to trade for hematite, the red mineral pigment so important to ritual observance and marking clan symbols on rock surfaces (pictographs).

Time and erosion have uncovered tools and utensils that were left by Puebloan peoples as they moved away from the canyon.

### THE PUEBLOAN EXIT

By the late 1100s something had tripped the switch on the climate box and once again the dry times and blowing sands were coming back. Stratigraphic studies of deposits along the river indicate a cluster of monumental spring floods in the 1170s, ironically followed by a decade of cruelly dry summers. Too much water, then not enough. This was the beginning of the end for the farmers and the end came quickly. The Colorado River began relentlessly stripping away the fertile terraces so essential to farming and the dry winds covered the gardens on the deltas with sand. After the catastrophic life-changing floods of the 1170s there were simply not enough organic-rich flood deposits each spring to counter the loss. The unabated erosion led to a critical drop in local water tables, further reducing the number of useful acres. Continued farming under drought conditions concentrated salts in the remaining soil and each successive harvest created less arable ground and more empty granaries.

By 1180 A.D. the cycle of increased rainfall that had precipitated the Puebloan flourish 150 years earlier had reversed itself and there was not enough rain to support a viable crop on the rims. The northwestern fringes of Pueblo civilization precariously situated at Grand Canyon were the first to fold under the early stages of the regional drought, which ultimately affected all of the farming people of the Southwest. Mesa Verde and

During Puebloan times, an alluvial terrace attached this bedrock bench and its group of four rooms to the river. Erosion after A.D. 1200 removed the sediment and left isolated blocks of limestone.

a few other culture centers would flourish for a while longer while the edges baked, but within a century Mesa Verde would be 90 percent abandoned as well.

As we know from earlier examples, such as the Altithermal drought of the Middle Archaic, this kind of change in the environment is not necessarily fatal to a culture, but it is directive. Once again, the locals were forced to make big decisions at the same time that decisions beyond their control were being made for them. This applied to all the inhabitants of canyon country, from the San Juan River 400 miles downstream through the Grand Wash Cliffs. The farther west in the canyon one travels, the harsher the climate and the quicker the drought would have had an effect. After two years of an extended drought, reserve and stored supplies of corn would be gone. Every succeeding crop would be more marginal and the carrying capacity of the land diminished in proportion. The wild plants and game they relied on as a dietary supplement were not close enough for the farmers. At their maximum population, the villages were looking into the face of a drought that would last longer than they could. Sadly, hunger stalked the villages. The oldest and the youngest would die first. If the babies cannot make it through the next winter and the young women are malnourished to the point that they are unable to conceive and bear children to term, then the culture is in deep deep trouble.

During this final Puebloan phase at Grand Canyon they constructed several thick-walled, seemingly defensive fort-like structures along the South Rim between Zuni Point and the Great Thumb. So there could well have been considerable tension and fear brought by the hunger accompanying the drought. Was the caution prompted by the ancestral Hualapai/Havasupai moving up-stream? Probably not. By advance parties of Southern Paiute on the North Rim? Quite possibly. Or other factors unknown or forgotten? Most likely. In any case all of the regional players were acting in response to drier times in an already dry place.

As times grew leaner it would have been easy to rationalize taking what was needed from those who had a surplus, unless there was a prohibition against such behavior. It was from these hard and dry times that the *A:Shiwi* (Zuni) developed the practice of never harming people who grow maize. This may be an insight as to how historical alliances were sorted out on the old western frontier. It was also during these trying times that ancestral Utes (Uto-Aztecan speakers) were moving north through the Plateau country into Utah and western Colorado as well as Navajo clans of *Dine'* (Athapaskan speakers) coming down the west slope of the Rocky Mountains, not to mention soon after Apache clans of *Inde'* (also Athapaskan speakers) coming from the east off of the Southern Plains, all putting even more pressure on an already complex situation. A Jemez oral tradition from these times tells of outlanders appearing out of nowhere with fires in the sky and men with sticks rolling burning skulls across the land—not unlike visions of the Vikings in Europe from the three previous centuries. By 1250 A.D. localized conflict and regional raiding were taking place across the American Southwest. Things were being sorted out and accelerated by regional climate change.

The manifestations of long-term drought were not only felt at the table, but they were also destructive to the cultural fabric beyond an individual's view. With the routine of daily life and the familiar cycle of the seasons disrupted, more time was devoted to gathering wild foods, leaving less time for the old chores and rituals. As the drought bore down, the springs that were crucial to facilitating long-distance trade across the low deserts dried up. The regional web of trade was broken, and ties of kinship and commerce outside the canyon were harder to maintain. The news carried on the lips of the traders was gone with the exotic goods they carried on their backs.

Even in the face of such hardship, life at Grand Canyon ground on, albeit scaled down.

**RETURNING FRAGMENTS OF THE PAST**

In 1990 the Native American Graves and Repatriation Act (NAGPRA) was signed into law in order to expedite the return of all native skeletal remains and grave goods to the appropriate tribe(s) and to foster a sense of respect for American Indian remains. "Prehistoric" human remains are rare at Grand Canyon. Literally, only a handful of skeletal materials have been found, telling us virtually nothing. The Havasupai and Hualapai cremated their dead, accounting for a lack of remains from that quarter. The prehistoric Pueblo generally disposed of the deceased by expedient burial in the trash midden outside the pueblo. High-status and special-circumstance burials are common to the prehistoric Puebloan world, but are yet to be uncovered in the canyon.

**HOPI TIES**

Twenty-three Hopi clans trace their origins to Grand Canyon. In particular, the Bear Strap, Greasy Eye Socket, and Bluebird clans migrated out of the Little Colorado gorge. They tell a tale of three groups of people leaving the canyon in turn. As they topped the rim, each group in succession encountered a dead bear in increasing states of decay. The first group out observed a rotting bear; the second, an all-but-decomposed one; the third saw bluebirds perched on the scattered bleached white bones. If oral tradition can be trusted to reflect a sequence of events, we can visualize these three clans coming out of the canyon in two years or less.

Tusayan Pueblo on the South Rim near Desert View was not built until after approximately 1180, maybe in response to the catastrophic flooding along the river. It was occupied at least through 1225 A.D., possibly as an outpost to keep an eye on their canyon. In the river gorge and on the South Rim we find small quantities of Pueblo ceramics made after 1250, so somebody was still around. There is no evidence of a single mass exodus from the canyon. Populations shrank the same way they had expanded in the eleventh century: quickly at first, followed by a slow dribble of departures. Some families undoubtedly left early, when it became apparent the drought would persist. Others stayed, readjusting in their own way. Small pockets of Virgin branch people isolated on the north side in the canyon's western reaches toughed it out, relying on the food they could gather, planting in wet nooks and crannies, having fewer children, going lighter on the calories. The real death blow to the old Puebloan cultures came three centuries later when the Spanish conquest brought European disease, enslavement, and Catholicism into the Southwest

Once again the wheel was turning. At some point farming at Grand Canyon became untenable and the final handful of Puebloan families moved out of the inner canyon. These people chose to leave their homes due to circumstances beyond their control, but not beyond their understanding. Too many people, not enough corn. They emerged into the broader world south of the canyon with an ominous drought still looming. So with their farming heritage and what goods they could carry, they walked out of the canyon into an increasingly dry and hostile world toward the Hopi Mesas and points south: the White Mountains, the Little Colorado River, and the well-watered Verde Valley. The Hopi Mesas acted as a refugium for corn-growing Uto-Aztecan speakers cast out on the landscape during the drought. These clans of Puebloans, along with clans from Fremont country to the north, some Virgin River people to the west, and survivors from Casas Grandes in Mexico, out of adversity, amalgamated over the next few centuries as the Hopi people.

Throughout the last millennium and into modern times the Hopi have maintained their ancient connections to the canyon, ritually in the kivas on the Hopi Mesas and physically by

Salt deposits along the Colorado River.

trekking into the canyon to collect salt and visit the *Sipapuni*, an elevated hot spring sacred to clans who directly trace their ancestry to the Grand Canyon. It represents their point of origin into the current world and their destination when they depart from it. Hot, iron-rich waters evaporate around the lip of the rounded opening, depositing a rind of earthy yellow pigment that is used in rituals.

The Hopi do not go into the canyon lightly; they prepare for a trip months in advance. Still farmers today, they leave nothing to chance. Several years ago I had the rare privilege of walking with a small group of Hopi Elders to this quiet place. These dedicated men said the world was dying. It was obvious to them and they were focused on reversing this trend through ceremony. After the observance, several of the men packed the thick yellow pigment into white tube socks. We walked the next few hours back to the boats and rowed to camp without saying much and over the next several days we did not speak of it again. This solemn act was part of the continuum running back more than nine centuries to the time when the ancestors of these men and their families were farmers in the canyon.

Even the drought did not stop this journey to the salt mine. It became an annual event amongst Hopi clans who had originated in the canyon and had migrated to Hopi. World War II truncated this 900-year-old trek when the federal government put the majority of Hopi medicine (holy) men in prison for refusing to serve in the war. The ability to collect salt was restored by the NPS in 1992. Since then the Hopi have again been making annual trips down the river and they stop at the salt mine. So the ritual was restored after a 50-year hiatus.

# STILL AT HOME IN THE CANYON: AN OLD SOLUTION TO AN OLD PROBLEM

As Puebloan populations dwindled at the end of the twelfth century, other cultures waiting in the wings were taking advantage of the vacuum. From the Mohave Desert in the far west end of the canyon the Yuman speaking Cerbat/Pai groups came to dominate the river corridor and rim on the south side all the way to Desert View. These are the direct ancestors of the Hualapai, Havasupai, and Yavapai People who are still in, on, and around Grand Canyon.

The Southern Paiute were already on the move, migrating southeastward from the Great Basin of eastern California and Nevada, stopping north of the Colorado River before the drought was in full swing. Archaeologically they are often called the Numic culture, because they speak a Uto-Aztecan language referred to as Numic. They have an ephemeral signature on the surface that is unique and because of that they are identifiable on the landscape. An early Numic/Southern Paiute site on the Colorado River system, but far upstream of Grand Canyon on the road to Dead Horse Point, near Moab, Utah dates to 1190 A.D., telling us that Paiutes could have easily been on the North Rim a decade or two before that, in random tandem with the drought.

## The Cerbat/Ancestral Pai

As the drought peaked, the Cerbat (the archaeological term)/ancestral Pai were well positioned geographically with 7,000-plus years of desert skills and foraging methods that would allow them to occupy the central canyon and flourish where the farmers could no longer be sustained. From their home territory to the west in Mata Widita, they had traded with the

Puebloans, but the archaeological record does not give up the exact day when they arrived in central Grand Canyon as permanent residents. Most likely after 1250 and before 1300. The Pai were moving up-canyon on the south bank in reaction to the same drought that was plaguing the Puebloan farmers and were undoubtedly on the move, one side canyon at a time, with families doing a slow hopscotch upstream as the vacuum grew, even before the complete Puebloan withdrawal.

A few researchers believe the Upland Pai pushed the prehistoric Pueblo out by force. This however is highly unlikely as the Pai groups on the south side of the river have remained friendly and on excellent terms with the Hopi, which probably reflects an old alliance from the drought times. Scattered conflict and raids were inevitable, but for the Puebloans the trouble was with the Paiutes on the north side, and other foreign groups, like Utes passing through the country and Athapaskan speakers knocking, unwanted, on the door, to the east. Not to mention the inevitable and often more intense internecine schisms that accompany hard times. It happened in other places in the Southwest, but we do not see it in the record at Grand Canyon. Conflict, if and when it took place, would have been on a limited scale. Small groups of people were drifting in and out of the region, not large armies. That, of course, makes the experience no less intense for the participants. It is also feasible that the few Puebloans who refused to leave home taught their ways of farming to the Pai and were absorbed. The Puebloans who left were not driven out at the tip of an arrow, but rather were prodded in the stomach by an empty spoon.

The Hualapai origin myth directs us to Mada Widita Canyon (Meriwhitica and Spencer on the modern maps) in the extreme west end of Grand Canyon. Their culture hero, Kathat Kanave, lived in Wahavo, a pueblo built in a cave high up on the canyon wall. He was an old man (and sometime coyote) who taught the Pai People the proper way to live in the main canyon. He told them what and when to plant, how to resolve disputes, what medicines could be found where and when, and other singularly Puebloan information—a pretty clear indication that some overlap existed between the two cultures. Mr. Kanave may well have been an old Virgin Puebloan man who wouldn't leave his home at Wahavo and stayed behind when his relatives packed up and left.

Early camps of the Cerbat/Pai would have looked much like this. The presence of tin cans is practically the only distinguishing feature of this Havasupai camp photographed by George Wharton James in 1900.

Cerbat/Pai and Cohonina archaeological sites are very different from Puebloan sites. Yet, in the canyon's west end there is a blending of the old and the new. Good flat places to live on are at a premium everywhere in the canyon and over the centuries a good spot often remains the only spot; so where it occurs, it is occupied. Artifacts blend together on the surface, causing anxiety for the archaeologist. Tizon Brownware pottery is a signature trait of the Cerbat and Upland Pai culture, originating at sites on the lower Colorado River and produced with little change between 1,200 and 250 years ago. The Pai manufactured it prior to their movement eastward into Grand Canyon proper. In fact sherds of Tizon Brownware have been found as far upstream as Unkar Delta on Puebloan sites dating to a century before the drought, well ahead of when the Puebloans split for greener pastures. Less than a century after the drought peaked, Hopi Yellow Wares start showing up on Hualapai sites along the river on the south bank. That fact alone makes a pretty good case for active trading and friendly continuity between the Puebloans and the Yuman speakers.

The Yuman speakers (Upland Pai) moved in small family groups in an established seasonal rhythm from water source to water source, hunting deer and bighorn sheep, gathering mesquite, sele seed, pinyon, prickly pear, their staple agave (mescal), and other plant foods. Barely discernible short-term camps would typically consist of few artifacts: a cleared circular area and rock ring where a *wuhbunvah*, or brush shelter, had stood, a small roasting pit, usually some hand-held tools, a grinding slab or anvil stone, a few scattered flakes, a handful of earth-brown Tizon sherds. Less often a Littleman point or the fragments of a prized Hopi tradeware bowl are present. Such a minimalist site represents a family on its traditional rounds, utilizing specific resources along the rims, benches, and river.

More complex long-term camps existed under the shelter of the rims and down along the river where side canyons open into the gorge. These were the home bases of extended families, where evidence is found of generational use over centuries: overlapping conical roasting pits 20 feet in diameter and 7 feet high, pictographs, digging sticks, broken pots, quids of chewed-and-spat-out agave fibers, all the debris of daily life that time has not engulfed. Common sites in western Grand Canyon include low rock walls next to game trails, used as blinds for hunting bighorn sheep.

All of the Upland Pai traded widely, exporting deer hides and dried meat, sheets of mescal, and highly valued red hematite (*gwat*) in buckskin bags, which moved eastward through the Hopi Mesas (in later times to the rich women in Spanish

Southwestern peoples used split green-wood tongs to collect the fruit of prickly pear cactus. Singeing the spines or using a rabbit-foot brush to rub them off, they ate the fruit fresh or split and dried it for storage. Tongs found at the mouth of National Canyon. Note burned tips. Shown at 80% of actual size.

Santa Fe) and downstream as far as Baja California, in exchange for Hopi and Mojave pottery, shell from the Gulf of California, and closer to our own time, cartridges, brass, coffee, and rifles.

By 1400 A.D. the Hualapai were the dominant group from the mouth of the Bill Williams River upstream along the South Rim to the confluence of the Little Colorado. Divided up into eleven or twelve geographically determined bands including the Havasupai, the Hualapai Confederation represented a people who spoke the same Yuman language, shared a heritage and an inherited landscape, and lived in what Dr. Robert Euler aptly described as *territorial equilibrium*.

In 1994 the National Park Service loaned me to the tribe at Peach Springs for a survey project, so I was out driving around on the Hualapai Reservation 150 miles west of Grand Canyon Village with the late tribal archaeologist Ron "Man" Susanyatame, a Hualapai descended from the Pine Springs band. He was a great conversationalist when provoked. When asked how he thought those families living in isolation down along the river kept in touch with the rest of their people, he didn't miss a beat: "It was the runners." The *Giv' iyam* (gov yem). "All they did was run. They carried the news … they got everything they needed from the People for bringing the news." Kept informed by the Giv' iyam, the families and bands could mobilize quickly against a common enemy or come together for rabbit drives and dances.

During the Big Drought the ancestral Havasupai (possibly descendants of the Cohonina) made strategic decisions based on local circumstances and headed along a track different from the other Hualapai bands. The Havasupai came to inhabit the permanently watered Cataract Canyon, one of Grand Canyon's largest side canyons. Because they were far enough west, where South Rim elevations are much lower and the winters milder than up-canyon, they wintered more comfortably in the sunnier uplands. After the agave was ripe in the spring, many families would move down in elevation, spending planting time through harvest in a number of suitable canyons draining off the South Rim—wherever the gardening was good and the summer sun less intense.

Before 1875 the Havasupai (*Nyava Kopai*) were the easternmost band of the Hualapai Confederation. Because they maintained gardens and lived half the year in Cataract Canyon, the U.S. Army considered them distinct from the Hualapai: "More like Hopis" was the Army mantra. Thus they were spared the brutal Hualapai War (1867–1872) and emerged into the twentieth century recognized by the federal government as a separate tribe.

The Hualapai call the Colorado River *Hakataia*. From the rim, the rapids and endless current seemed to them to be alive like an animal, the rapids its backbone, or *haitat*. From a bird's-eye view it does not take much imagination to see that the rapids look like an undulating spine. Today the Hualapai and Havasupai Reservations extend only from Cataract Canyon to Lake Mead, occupying the forested rim country down to the river for more than a hundred miles. In the late nineteenth century they were pushed to the edge of this lean country, not this time by a change in the weather, but by individuals who were greedy for gold and America's need, particularly in this case the railroad's need for land, water, and timber.

## THE SOUTHERN PAIUTE

Another car on the north-south Uto-Aztecan culture train arrived at Grand Canyon during the twelfth and thirteenth centuries, just in time for the drought. Paiute hunter-gatherers entered into a landscape on the north side of the Colorado River that was sparsely populated by Virgin Puebloan families. Here and there isolated farmers were still hanging on, as at Pinenut Pueblo, and possibly the settlements around Mount Trumbull. It is probably from these residual groups of colonial Puebloans that the first wave of Paiutes learned how to supplement their wild foods with corn and squash grown around the springs and down the side canyons. On their earliest trips into the inner canyon they would have encountered farmsteads virtually intact, with empty ollas on the floors of the abandoned rooms and occasional volunteer corn and squash plants in the untended plots.

Southern Paiute and Cerbat/Pai sites are often hard to differentiate based solely on artifacts. A rule of thumb for the Grand Canyon is "Paiute north (right) bank, Cerbat/Pai south (left) bank," but this only works in general. Like the Pueblo groups before them, the Southern Paiute utilized both sides of the river as necessity dictated. During the mid-1800s, Mormon settlement of the Arizona Strip and North Rim forced Southern Paiute bands to seek shelter across the river with their neighbors. Before the Mormons came, the canyon afforded protection from Mountain Ute, Navajo, and Spanish slavers who roamed and raided in the region. Similarly, the government's war against the Hualapai (1867–1872) drove families and bands deeper into the canyon and occasionally north across the river onto Paiute ground.

The distinctive style of Paiute pictographs is evident throughout the Grand Canyon river corridor.

In 1869 explorer John Wesley Powell and his hungry men pillaged the unattended Paiute gardens along the river to enhance their pathetic diet of rancid bacon, flat biscuits, and black coffee. The immature greens made the entire crew violently ill.

To the Southern Paiute, the entire Grand Canyon country is a land of power. The specific name of the canyon is *Pipaxa' uipi*, Big River Canyon. They are great singers and in song they describe and commemorate their relationship to the Canyon. Songs are differentiated into those that pertain to the trails of the living and those that individuals sing to guide themselves on the trail to the afterlife. Like the Hualapai, the Southern Paiute had a guild of runners who carried messages from camp to camp, except that Paiute messages were coded into knotted strings called *tapitcapi*.

The Southern Paiute cultural landscape was held together by a complex system of trails connecting the far-flung water sources in Grand Canyon and beyond into Nevada, Utah, and southern California. The web of trails spreading out across the surface of the Paiute world was committed to memory by song. Because of summer heat or the need to deliver an urgent message, runners moved confidently through the night guided by a song that had been passed down orally through generations. This information, gathered from modern Paiute people, illustrates how limiting it can be to explain an intricate way of life by just studying

A variety of pendants and worked stone from Grand Canyon.

## MUSIC: SONGS, DANCE, PERCUSSION, AND THE RHYTHMS OF LIFE

There is nothing more intrinsic to humanity—aside from language, fire, and the opposable thumb—than music: our voices, rhythms, and ritual motions, collectively or individually. These are rooted experiences for us all from birth to death. This is as true today as it was 1,000 or 20,000 years ago and remains so for modern native peoples. All tribes had their own songs, both individual and communal, for every occasion: individual power, group focus, recounting events, births and deaths, before hunts and after, puberty rights, and spontaneous celebration. The same applies to dance—ritual, social, or to help make it rain—generally in a clockwise, circular motion to reflect the cycles of seasons and the track of the sun and the moon across the sky. Social dancing has always been equally popular, often a two-step in opposing lines, "coming together and going apart," all held together by the rhythm of drums, gourd rattles, and voices. We all enjoy one or all of these very human inventions and behaviors, which are culturally specific, but universally practiced. Vincent Randall, a Dilzhe'e (Western Apache) singer and tribal elder, says the monotonous "Dumdumdumdum Dumdumdumdum of Indian drumming in cartoons and Westerns is pure fiction. True "Indian drumming is bum-BUM-bum-BUM-bum-BUM!" The beat of the human heart.

**A SOURCE OF POWER**

For American Indians, there is an important connection between spiritual power and place. The Southern Paiute call sites of traditional significance *pohaghani*, houses of power. At Grand Canyon there are numerous examples associated with several cultures: the Sipapuni, Ribbon Falls, the Salt Cave, Deer Creek Falls, Three Springs, and countless others, forgotten or not spoken of.

Several aspects of the power that resides in the natural world come together at the mouth of Prospect Canyon, the site of Lava Falls Rapid. Here the transition from high desert to low desert becomes apparent. Along the river, vegetation changes to ocotillo, barrel cactus, and mesquite. The rim elevations drop. The ponderosa pine forest gives way to juniper breaks, groves of stunted oak, sage flats, and expanses of barren rock. Dormant volcanoes and cinder cones protrude, and deeply incised canyons carve into the uplands.

At least seven times before humans arrived at Grand Canyon, lava cascaded over the rim at this point and flowed, steaming, down the river channel, cooling into natural dams that pooled the river into large, elongated lakes for thousands of years at a stretch. Each time, the river wore most of the lava away, but some remained locked to the canyon walls, creating towering jointed cliffs.

In the middle of the Colorado River a mile above Lava Falls, a residual plug of basalt creates an imposing black island which today is known as Vulcans Anvil. The Hualapai call it *wi-nyak-da-lupa*, the big black rock. They, along with the Havasupai and the Southern Paiute, consider the big black rock and surrounding landscape sacred.

*"It is easy to get lost in the details of material culture and archaeological remains with the consequence that the over whelming presence of the landscape fades into the background."*
—Trevor Garnham 2004

broken pottery, hand tools, and projectile points. When their system of organized communication is applied to the past, it adds a little flesh and motion to the texture of Paiute society.

As with their predecessors and contemporary neighbors, rock art was an important form of expression to the Paiute. One term the Paiute use is *tumpituxwinap*, inadequately translated as storied rocks. Another term is *tumpim po 'okant*, roughly meaning having a lot to tell on the rocks. When the glyphs and graphs cannot be deciphered because the symbols come from another culture or are the work of a shaman, they are referred to as *navastump*, nonsense rock. It is assumed that the symbols have some specific meaning, but

A selection of projectile points from Grand Canyon.

that meaning is lost to the viewer. One of the last Paiute pictograph panels to be executed in Grand Canyon details a Ghost Dance held down along the river in the 1880s. This religious movement was the inspiration of Wovoka, a Northern Paiute man who, after becoming discouraged with Christianity, envisioned (in a spate of wishful thinking) the immediate demise of Euro-American culture and the disappearance of the federal government brought about by repeated staging of the Ghost Dance by all American Indians. Sadly, the movement culminated in the tragic massacre at Wounded Knee, South Dakota in the winter of 1890.

The Southern Paiute efficiently gleaned a living from spare land. It was not a shift in the climate or ecological catastrophe that pushed them out of the canyon, but rather the expansion of European-American culture, specifically cattle culture, into the region from 1860 to 1885. A lifestyle that had existed for more than 700 years in a true balance with the available resources was snuffed out in a single generation. Several hundred archaeological sites at Grand Canyon mark its passing. Modern Southern Paiute people live on patchwork Reservation lands north of Grand Canyon proper in Arizona, Utah, and Nevada.

## EPILOGUE: 1540 SPANIARDS ON THE South Rim

A mere 39 years after Columbus' initial voyage, the first and maybe the most incredible journey by Europeans through the American Southwest was made by Cabeza de Vaca and his five comrades in 1531–1536 across the southern edge of what would become Texas, New Mexico, and Arizona. Cabeza was one of just a handful of men out of over a thousand to survive the failed Narvaez expedition from Cuba to Florida in 1529. After surviving captivity on the Gulf Coast, Cabeza, four other Spanish officers, and Esteban, a captive Moor, made their way west on foot. After five years of trekking across this uncharted land they were recaptured by Spanish slavers in northern Sonora. Broke, out of favor with the Crown, and lacking friends, the men were dead-ended. The officer who owned Esteban sold him to Marcos de Niza, an Italian priest in Mexico City, to pay for passage back to Spain.

Esteban—who was brave, literate, and resourceful—told de Niza that while in the wilds of the far north they had seen cities paved in gold, as many as seven—grander than anything Cortez had seen in the Valley of Mexico or Pizzaro in Peru. Based on this tall-tale de Niza easily solicited funding from the greed-driven Church and headed north with Esteban and a handful of full-blooded Aztec novitiates (that's how early this was). It was a long and dangerous trip. Fast forward one year: de Niza halted several miles from Zuni and sent Esteban and some of the Aztecs ahead to parley. He was immediately taken captive and thrown into a turkey pen outside the walled gate. The next morning Esteban was filled full of arrows and killed by the Zuni. The young Aztecs were told to go home and never come back. De Niza along with the converted Aztecs hightailed it back to Mexico City, where a tale of failure might land him in chains on a ship back to Spain or worse. So like any self-serving explorer on somebody else's dime, he lied. "Yes," the Cities of Gold, El Cibola, they are there, he told them. The Church and the royal officials asked if he had gone to these cities? "No," they had been ambushed and were forced to flee, but he had seen the golden roofs and domes from afar, "gleaming in the sun."

In 1537 this tale was all the buzz in New Spain and all it took to hook the ambitious young governor of New Galicia, Francisco de Bautista de Ebarra de Montoya i Maria del Gomez Coronado. Governor Coronado lobbied hard for the money, but more importantly for the Imperial Charter that made it legal to take an expedition north. To do otherwise was to be an outlaw. A modern economist has calculated that mounting the expedition to Cibola cost the Crown an amount approximately comparable to the Apollo Space Program. Coronado's spurs were on the line. The expedition included 300 noblemen from all over Europe, adventurers, herdsmen, blacksmiths, camp followers, thousands of sheep, cattle, and horses along with portable forges, bakeries, trade goods and a variety of new European diseases as passengers. The expectations were high. By the summer of 1540 Coronado and his noisy column were camped outside of Cibola (Zuni). They were not happy. No gold, no silver, no riches of the Spanish stripe, "only a few turquoises and cotton blankets." This did not bode well for the governor, and to pursue every avenue of success he sent out scouting parties in all directions in the hope of finding something worthwhile.

He sent one of his commanders, Pedro de Tovar, to the northwest, and with Zuni guides he came to Hopi, where he learned of a great river running off to the west. Tovar returned to base camp at Cibola with his findings. Coronado then sent Maestro de Campo Juan de Cardenas and twelve other men out again to find this river and the gold it might contain. After coercing guides at Hopi he and his men were led across the dry country between the Mesas and the Little Colorado River. Ultimately they found themselves on the South Rim somewhere near Desert View. Thus in early September of 1540 Europeans peered over the edge of Grand Canyon for the first time. They decided to climb down. The Hopi had wisely taken them to a place where no trails were visible from the rim. Undeterred by the lack of a direct path (they were conquistadors after all!), a captain Melagosa, Juan Galenos, and an anonymous soldier claimed to have made the descent about a third of the way down, but it was probably much less. Overcome by exhaustion and lack of water they picked their way back out and reported to Cardenas that the rocks, which looked to be the size of a man from above, were taller than the "Tower of Seville," and the river that looked to be only several feet across from the rim was actually a raging torrent. They had no point of reference as "this Canyon" was completely off their little European scale.

There was no water for the big Arab horses the Spaniards rode and the grain they ate was gone. So in their heavy armor on their thirsty horses they left and went back to Coronado's melancholy camp at Cibola. The Spaniards never returned to the canyon. The Hopi guides had told Cardenas, because of their long experience, that when they traveled through this country they brought women carrying gourds full of water that they buried at the halfway point for the return trip. The Hopi had known this for centuries.

In 1776, on a trek north from Tucson, Father Garces encountered the Grand Canyon and spent five days with the Havasupai in Cataract Canyon. He could see campfires on the distant North Rim and the Havasupai told him they were the camps of *Payuces* (Southern Paiute). He noted in his journal, "I am astonished at the roughness of the country," and he left. Later that same year two priests, Dominguez and Escalante, left Santa Fe on a mission to harvest Native souls. After a circuitous route through western Colorado and north-central Utah they ended up on the north side of the Colorado River in Grand Canyon at the mouth of the Paria River. The river could not be crossed. They named the camp "Get Out if You Can," and later crossed upstream at what came to be known as the "Crossing of the Fathers," now under several hundred feet of water in Lake Powell. They slow-walked their way back to Santa Fe and never saw the "horrible abyss" again.

The Spaniards left no significant imprint on Grand Canyon, except for the name of the river itself, COLORADO. Nonetheless there was not enough water in the right places and not enough gold in the rocks to make it worth staying. Just like the old Puebloans, they refocused their efforts on the better-watered highlands in New Mexico—much to the detriment of the Puebloans. Today there is a lovely hotel on the South Rim named after Pedro Tovar where it is no problem getting a cool drink on a hot day. Maestro de Campo Cardenas, butcher of Tiguex Pueblo in New Mexico and leader of the first European expedition to the Grand Canyon, has a butte and a creek named after him. Cardenas Creek is generally dry, but occasionally flows into the river near mile 71 on the left bank. There is an old Pueblo farmstead there buried in the sand dunes, and overlooking those dunes is the ancient "Hilltop Fortress."

# CONCLUSION

Although the past will always be ephemeral, it remains doggedly with us. The fragments of individual lives, families, and their greater cultures lie scattered on the surface and buried within Grand Canyon. Those broken pots, dormant tools, concentrations of charcoal, and hand-drawn symbols allow us a tantalizing, mist-covered portal into history. Archaeological sites rarely preserve a moment frozen in time. More correctly, sites can represent a distinct episode or a composite of days, seasons, or years of activity, partially preserved and thus partially understood.

From studying archaeological sites, we can conclude this much: Whether you are a cactus, a lizard, a rock, or a person, the Grand Canyon is unforgiving. The more reliance a culture places on altering the environment to maintain its standard of living, the shorter is its story. There is not a shred of compromise built into the system. Like the geological history of the earth revealed in the walls of the canyon, the various cultures exist in a greater context. Complicated further by the intangible nuances of the human mind, the sites represent a great deal more than artifacts can express.

During the 1930s, by total coincidence, a group of young Paiutes from the North Rim and a group of young Hualapai from the South Rim simultaneously worked their way down to the basalt-lined benches overlooking Lava Falls, where, separated by the river, each group built a large fire and celebrated. They yelled back and forth across the divide, but nobody could make out what was said and nothing else came of it. Forty years later at a more official function, a Paiute woman began talking about the Grand Canyon, its importance to her people, and the bonfire she attended. In the assembled group was a Hualapai woman who had been dancing on the opposite bank. It was an unexpected reunion. The water flowing through Grand Canyon is still dividing and bringing people together.

The Archaic hunter, the Pueblo family, or a Hualapai tracker from just two centuries ago would recognize the place, but probably not be too happy about the crowds. The Grand Canyon is a different entity now. It is a modern place and a managed space. The rock is still the same, but the cultural landscape is nothing like it was. Yet now as then it is still all about the water. The National Park Service does a truly magnificent job with limited funding, but it's an uphill battle to properly manage all of the visitors and multiple resources simultaneously. The Forest Service manages the forested rims back to the Park boundary and the Bureau of Reclamation manages the water that flows in the river through the canyon.

Today the Colorado River in Grand Canyon is essentially a public energy and watering trough with adjustable spigots situated at each end up and downstream. Many agencies, municipalities, corporations, and individuals have a straw stuck in it and that will not diminish any time soon. Ultimately, whatever may happen on the human scale, in the end the Canyon heals itself.

# index

## ACKNOWLEDGEMENTS

All projects great and small are, behind the scenes, a group effort. To my parents and grandparents, who encouraged me to swim in the river of culture and science. To my wife, Marcelle, for her support. To the people who taught me how to be a field archaeologist: Mark Bond, Debbie Westfall, Bill Davis, Steve Kalasz, and Richard Carrillo (because I'm really a geologist). My admiration to all of the field personnel on the epic River Corridor Survey of 1990-1991, especially the original D-Crew, Victoria Clark-Davenport and Jonathan Till, and Helen Fairley, who held it all together. To Jan Balsom, who had our backs high above on the South Rim. A special thanks to the boatmen and cooks who got me safely home from more than 50 trips: Deb Petersen, John Davenport, Ed "Participate" Smith, Lisa Whisnant, Bob Grusy, Nancy "Gidget" Rivers, John Hirsch, Kenton Grua, Brian Dierker, Kim Crumbo, Kelly Smith, Jim Traub, Roger Henderson, Bruce Keller, Tamara Wiggans, Chris Geanious, Nancy McCleskey, Dave Derosiers, Dick Clark, Thad Stewart, Martha Clark-Stewart, Niki Dunn LeClaire, Denise Napalitano, and Doug Deutschlander.. To NPS staff Mike Quinn, Coleen Hyde, and Kim Besom for all their help. For their eyes on the words: Duane Hubbard, Elizabeth Brennan, and Jen Kunde-Dierker. To our editor, Louella Holter. To my friends and colleagues, who in various ways added to the project: Joe Pachak, Wendy Hodgson, Ron Man Susanyateme, Christine Göetze, Ivo Luchitta, Angie Bullets, Leigh Kuwanwisima, Loretta Jackson, Rex Tilousi, Vincent E. Randall, Elizabeth Smith-Rocha, Victor Hiawatha Smith, David Sine, Lyn Murray, Mark Taylor, J. E. Whipp, Jr., The Boyz from Illinoiz Canyoneers Society and my GRCA hiking partner Pat "The Underminer" Harkins. A special thanks to our graphic designer, Rudy Ramos, my publisher, Lillian "Lulu" Santamaria, and to Addie for getting us across the finish line. Thanks to everyone.

See you all downstream!
CMC

## ABOUT THE AUTHOR

Chris Coder has worked as a freelance archaeologist on the Colorado Plateau and Great Plains since 1980. He earned a Bachelor's degree in Geology from Augustana College, Rock Island, Illinois (1977) and a Master's in Anthropology from Northern Arizona University (1983). He began his Grand Canyon experience as a backpacker in 1974, and spent 1990–1997 as an NPS project archaeologist in the river corridor. Chris currently lectures and writes, and since 1997 has been the archaeologist for the Yavapai-Apache Nation. He lives with his wife Marcelle outside of Flagstaff, Arizona.

Photo taken at the Isle of Mull, Scotland.

## SUGGESTED READING

Anderson, K., Kranzler, B., Neff, T., Nez, J., Smith, S., and Spurr, K. *Grand Archaeology, Excavation and Discover along the Colorado River.* 2012, Volume 7, Number 1, Plateau. Museum of Northern Arizona.

Basso, Keith H. *Wisdom sits in places*. 1996. University of New Mexico Press, Albuquerque.

Cordell, Linda S. and George J. Gumerman. editors. *Dynamics of Southwest Prehistory*. 1989. Smithsonian Institution Press.

Fontana, Bernard L.. *Entrada. The Legacy of Spain and Mexico in the United States*. 1994. Southwest Parks and Monuments Association.

Garnham, Trevor. *Lines on the Landscape Circles from the Sky*. 2004. Tempus Publishing, Stroud, England.

Hirst, Stephen and the Havasupai Tribe. *I Am the Grand Canyon: Story of the Havasupai People.* 2006. Grand Canyon Conservancy.

Plog, Stephen. *Ancient Peoples of the American Southwest*. 1997. Thames & Hudson.

Riley, Carroll L .. *Rio del Norte: People of the Upper Rio Grand from Earliest Times to the Pueblo Revolt*. 1995. University or Utah Press.

Schwartz, Douglas W.. *On the Edge of Splendor: Exploring Grand Canyon's Human Past*. 1989. School or American Research.

Smiley, Frances E., Christian E. Downum and Susan G. Smiley. *The Archaeology of Grand Canyon: Ancient Peoples, Ancient Places*. 2017. Grand Canyon Conservancy.

Walker, Steven L. *The Southwest: A Pictorial History of the Land and its People*. 1993, Camelback Design Group and Canyonlands Publications.

## PHOTOGRAPHY CREDITS

Kim Besom 26, 27t left; Mike Buchheit 10; Chris Coder 27m and b left, 27b right, 54, 62; Marcelle Coder 71; Dawn Kish 8b; Lois Ellen Frank 29; George H. H. Huey 40 sherds; Gary Ladd 27t right, 46, 51; Museum of Northern Arizona 38, NA6887.15; NPS 9, 12t, 21, 24, 39, 40 (pot), 49, 59; NPS Michael Quinn 1, 3, 5, 9, 12b, 13, 23, 25, 49 both, 51, 60, 67, 69; Northern Arizona University, Cline Library 8t, NAU.PH.98.70; Courtesy of the Peabody Museum of Archaeology and Ethnology, Harvard University, 2004.24.5833, 28; School of American Research, Douglas W. Schwartz 37, 57; Shutterstock 1129062788 19; Utah Museum of Natural History, University of Utah, Laurel Casjens 33 both; Lisa Whisnant 53.

## ILLUSTRATION CREDITS

Roy Anderson, Daily Life at Tusayan Pueblo, 36, courtesy of the artist; Chris Coder 35, 45, 47, 52; Susah Leach 39, 63; NPS 6, 26-27, 46; Joe Pachak 21 (points), 24, 30; School of American Research 9 (site plan, Bright Angel); Thames & Hudson Ltd; 23. iStock.com 35 Agave, 36 Samarai, 60 Prickly pear.

*t=top; b=bottom; m=middle; of page Individual photographers and artists hold copyrights to their work.*